Teacher Edition

Eureka Math
Grade 4
Module 7

Special thanks go to the Gordon A. Cain Center and to the Department of Mathematics at Louisiana State University for their support in the development of *Eureka Math*.

For a free *Eureka Math* Teacher Resource Pack, Parent Tip Sheets, and more please visit www.Eureka.tools

Published by the non-profit Great Minds

Copyright © 2015 Great Minds. No part of this work may be reproduced, sold, or commercialized, in whole or in part, without written permission from Great Minds. Non-commercial use is licensed pursuant to a Creative Commons Attribution-NonCommercial-ShareAlike 4.0 license; for more information, go to http://greatminds.net/maps/math/copyright. "Great Minds" and "Eureka Math" are registered trademarks of Great Minds.

Printed in the U.S.A.
This book may be purchased from the publisher at eureka-math.org
10 9 8 7 6
ISBN 978-1-63255-376-8

***Eureka Math: A Story of Units* Contributors**

Katrina Abdussalaam, Curriculum Writer
Tiah Alphonso, Program Manager—Curriculum Production
Kelly Alsup, Lead Writer / Editor, Grade 4
Catriona Anderson, Program Manager—Implementation Support
Debbie Andorka-Aceves, Curriculum Writer
Eric Angel, Curriculum Writer
Leslie Arceneaux, Lead Writer / Editor, Grade 5
Kate McGill Austin, Lead Writer / Editor, Grades PreK–K
Adam Baker, Lead Writer / Editor, Grade 5
Scott Baldridge, Lead Mathematician and Lead Curriculum Writer
Beth Barnes, Curriculum Writer
Bonnie Bergstresser, Math Auditor
Bill Davidson, Fluency Specialist
Jill Diniz, Program Director
Nancy Diorio, Curriculum Writer
Nancy Doorey, Assessment Advisor
Lacy Endo-Peery, Lead Writer / Editor, Grades PreK–K
Ana Estela, Curriculum Writer
Lessa Faltermann, Math Auditor
Janice Fan, Curriculum Writer
Ellen Fort, Math Auditor
Peggy Golden, Curriculum Writer
Maria Gomes, Pre-Kindergarten Practitioner
Pam Goodner, Curriculum Writer
Greg Gorman, Curriculum Writer
Melanie Gutierrez, Curriculum Writer
Bob Hollister, Math Auditor
Kelley Isinger, Curriculum Writer
Nuhad Jamal, Curriculum Writer
Mary Jones, Lead Writer / Editor, Grade 4
Halle Kananak, Curriculum Writer
Susan Lee, Lead Writer / Editor, Grade 3
Jennifer Loftin, Program Manager—Professional Development
Soo Jin Lu, Curriculum Writer
Nell McAnelly, Project Director

Ben McCarty, Lead Mathematician / Editor, PreK–5
Stacie McClintock, Document Production Manager
Cristina Metcalf, Lead Writer / Editor, Grade 3
Susan Midlarsky, Curriculum Writer
Pat Mohr, Curriculum Writer
Sarah Oyler, Document Coordinator
Victoria Peacock, Curriculum Writer
Jenny Petrosino, Curriculum Writer
Terrie Poehl, Math Auditor
Robin Ramos, Lead Curriculum Writer / Editor, PreK–5
Kristen Riedel, Math Audit Team Lead
Cecilia Rudzitis, Curriculum Writer
Tricia Salerno, Curriculum Writer
Chris Sarlo, Curriculum Writer
Ann Rose Sentoro, Curriculum Writer
Colleen Sheeron, Lead Writer / Editor, Grade 2
Gail Smith, Curriculum Writer
Shelley Snow, Curriculum Writer
Robyn Sorenson, Math Auditor
Kelly Spinks, Curriculum Writer
Marianne Strayton, Lead Writer / Editor, Grade 1
Theresa Streeter, Math Auditor
Lily Talcott, Curriculum Writer
Kevin Tougher, Curriculum Writer
Saffron VanGalder, Lead Writer / Editor, Grade 3
Lisa Watts-Lawton, Lead Writer / Editor, Grade 2
Erin Wheeler, Curriculum Writer
MaryJo Wieland, Curriculum Writer
Allison Witcraft, Math Auditor
Jessa Woods, Curriculum Writer
Hae Jung Yang, Lead Writer / Editor, Grade 1

Board of Trustees

Lynne Munson, President and Executive Director of Great Minds
Nell McAnelly, Chairman, Co-Director Emeritus of the Gordon A. Cain Center for STEM Literacy at Louisiana State University
William Kelly, Treasurer, Co-Founder and CEO at ReelDx
Jason Griffiths, Secretary, Director of Programs at the National Academy of Advanced Teacher Education
Pascal Forgione, Former Executive Director of the Center on K-12 Assessment and Performance Management at ETS
Lorraine Griffith, Title I Reading Specialist at West Buncombe Elementary School in Asheville, North Carolina
Bill Honig, President of the Consortium on Reading Excellence (CORE)
Richard Kessler, Executive Dean of Mannes College the New School for Music
Chi Kim, Former Superintendent, Ross School District
Karen LeFever, Executive Vice President and Chief Development Officer at ChanceLight Behavioral Health and Education
Maria Neira, Former Vice President, New York State United Teachers

This page intentionally left blank

A STORY OF UNITS

GRADE 4

Mathematics Curriculum

GRADE 4 • MODULE 7

Table of Contents

GRADE 4 • MODULE 7

Exploring Measurement with Multiplication

Module Overview 2

Topic A: Measurement Conversion Tables 10

Topic B: Problem Solving with Measurement 79

Topic C: Investigation of Measurements Expressed as Mixed Numbers 153

End-of-Module Assessment and Rubric 189

Topic D: Year in Review 198

Answer Key 243

OVERVIEW

In this module, students build their competencies in measurement as they relate multiplication to the conversion of measurement units. Throughout the module, students explore multiple strategies for solving measurement problems involving unit conversion.

In Topic A, students build on their work in Module 2 with measurement conversions. Working heavily in customary units, students use two-column conversion tables (**4.MD.1**) to practice conversion rates. For example, following a discovery activity where students learn that 16 ounces make 1 pound, students generate a two-column conversion table listing the number of ounces in 1 to 10 pounds. Tables for other measurement units are then generated in a similar fashion. Students then reason about why they do not need to complete the tables beyond 10 of the larger units. They use their multiplication skills from Module 3 to complete the tables and are able to see and explain connections such as (13 × 16) = (10 × 16) + (3 × 16). One student could reason, for example, that, "Since the table shows that there are 160 ounces in 10 pounds and 48 ounces in 3 pounds, I can add them together to tell that there are 208 ounces in 13 pounds." Another student might reason that, "Since there are 16 ounces in each pound, I can use the rule of the table and multiply 13 pounds by 16 to find that there are 208 ounces in 13 pounds."

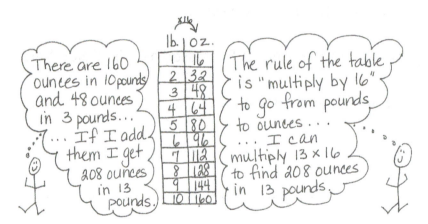

As the topic progresses, students solve multiplicative comparison word problems. They are then challenged to create and solve their own word problems and to critique the reasoning of their peers (**4.OA.1**, **4.OA.2**). They share their solution strategies and original problems within small groups, as well as share and critique the problem-solving strategies used by their peers. Through the use of guided questions, students discuss not only how the problems were solved, but also the advantages and disadvantages of using each strategy. They further discuss what makes one strategy more efficient than another. By the end of Topic A, students have started to internalize the conversion rates through fluency exercises and continued practice.

Topic B builds upon the conversion work from Topic A to add and subtract mixed units of capacity, length, weight, and time. Working with metric and customary units, students add like units, making comparisons to adding like fractional units, further establishing the importance of deeply understanding the unit. Just as 2 fourths + 3 fourths = 5 fourths, so does 2 quarts + 3 quarts = 5 quarts. 5 fourths can be decomposed into 1 one 1 fourth, and therefore, 5 quarts can be decomposed into 1 gallon 1 quart. Students realize the same situation occurs in subtraction. Just as $1 - \frac{3}{4}$ must be renamed to $\frac{4}{4} - \frac{3}{4}$ so that the units are alike, students must also rename units of measurements to make like units (1 quart − 3 cups = 4 cups − 3 cups). Students go on to add and subtract mixed units of measurements, finding multiple solution strategies, similar to the mixed number work in fractions. With a focus on measurement units of capacity, length, weight, and time, students apply this work to solve multi-step word problems.

I can rename 8 quarts so I have enough cups to subtract!

$$8qt\ 1c - 6qt\ 3c = 7qt\ 5c - 6qt\ 3c = 1qt\ 2c$$
$$7qt\ 4c$$

In Topic C, students reason how to convert larger units of measurements with fractional parts into smaller units by using hands-on measurements. For example, students convert $3\frac{1}{4}$ feet to inches by first finding the number of inches in $\frac{1}{4}$ foot. They partition a length of 1 foot into 4 equal parts and find that $\frac{1}{4}$ foot = 3 inches. They then convert 3 feet to 36 inches and add 3 inches to find that $3\frac{1}{4}$ feet = 39 inches. This work is directly analogous to earlier work with fraction equivalence using the tape diagram, area model, and number line in Topics A, B, and D of Module 5. Students partitioned a whole into 4 equal parts, decomposed 1 part into 3 smaller units, and found 1 fourth to be equal to 3 twelfths. The foot ruler is partitioned with precisely the same reasoning. Students close the topic by using measurements to solve multi-step word problems that require converting larger units into smaller units.

The End-of-Module Assessment follows Topic C.

Students review their year in Topic D by practicing the skills they have learned throughout the modules. Additionally, they create a take-home summer folder. The cover of the folder is transformed into the student's own miniature personal white board, and a collection of activities from the lessons within this topic are placed inside the folder to be practiced throughout the summer. Students practice major skills and concepts learned throughout the year in these final four lessons, including measuring angles and drawing lines, multiplication and division, and addition and subtraction through guided group work, fluency activities, and vocabulary games.

Notes on Pacing for Differentiation

Module 7 affords students the opportunity to use all that they have learned throughout Grade 4 as they first relate multiplication to the conversion of measurement units and then explore multiple strategies for solving measurement problems involving unit conversion. Module 7 ends with practice of the major skills and concepts of the grade as well as the preparation of a take-home summer folder. Therefore, it is not recommended to omit any lessons from Module 7.

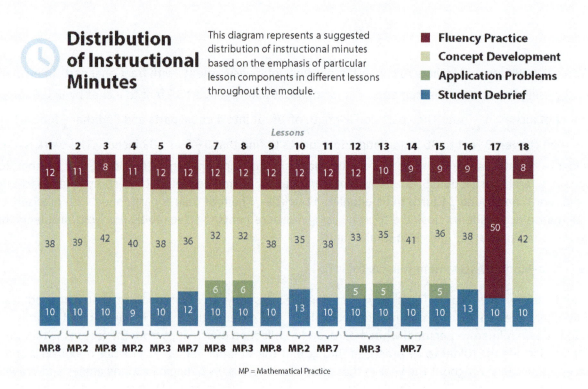

Focus Grade Level Standards

Use the four operations with whole numbers to solve problems.

4.OA.1 Interpret a multiplication equation as a comparison, e.g., interpret 35 = 5 × 7 as a statement that 35 is 5 times as many as 7 and 7 times as many as 5. Represent verbal statements of multiplicative comparisons as multiplication equations.

4.OA.2 Multiply or divide to solve word problems involving multiplicative comparison, e.g., by using drawings and equations with a symbol for the unknown number to represent the problem, distinguishing multiplicative comparison from additive comparison. (See CCSS-M Glossary, Table 2.)

4.OA.3	Solve multi-step word problems posed with whole numbers and having whole-number answers using the four operations, including problems in which remainders must be interpreted. Represent these problems using equations with a letter standing for the unknown quantity. Assess the reasonableness of answers using mental computation and estimation strategies including rounding.

Solve problems involving measurement and conversion of measurements from a larger unit to a smaller unit.[1]

4.MD.1	Know relative sizes of measurement units within one system of units including km, m, cm; kg, g; lb, oz.; l, ml; hr, min, sec. Within a single system of measurement, express measurements in a larger unit in terms of a smaller unit. Record measurement equivalents in a two-column table. *For example, know that 1 ft is 12 times as long as 1 in. Express the length of a 4 ft snake as 48 in. Generate a conversion table for feet and inches listing the number pairs (1, 12), (2, 24), (3, 36), …*
4.MD.2	Use the four operations to solve word problems involving distances, intervals of time, liquid volumes, masses of objects, and money, including problems involving simple fractions or decimals, and problems that require expressing measurements given in a larger unit in terms of a smaller unit. Represent measurement quantities using diagrams such as number line diagrams that feature a measurement scale.

Foundational Standards

3.OA.1	Interpret products of whole numbers, e.g., interpret 5 × 7 as the total number of objects in 5 groups of 7 objects each. *For example, describe a context in which a total number of objects can be expressed as 5 × 7.*
3.OA.3	Use multiplication and division within 100 to solve word problems in situations involving equal groups, arrays, and measurement quantities, e.g., by using drawings and equations with a symbol for the unknown number to represent the problem.
3.OA.5	Apply properties of operations as strategies to multiply and divide. *Examples: If 6 × 4 = 24 is known, then 4 × 6 = 24 is also known. (Commutative property of multiplication.) 3 × 5 × 2 can be found by 3 × 5 = 15, then 15 × 2 = 30, or by 5 × 2 = 10, then 3 × 10 = 30. (Associative property of multiplication.) Knowing that 8 × 5 = 40 and 8 × 2 = 16, one can find 8 × 7 as 8 × (5 + 2) = (8 × 5) + (8 × 2) = 40 + 16 = 56. (Distributive property.)*
3.OA.7	Fluently multiply and divide within 100, using strategies such as the relationship between multiplication and division (e.g., knowing that 8 × 5 = 40, one knows 40 ÷ 5 = 8) or properties of operations. By the end of Grade 3, know from memory all products of two one-digit numbers.
3.NBT.3	Multiply one-digit whole numbers by multiples of 10 in the range 10–90 (e.g., 9 × 80, 5 × 60) using strategies based on place value and properties of operations.

---

[1]The focus now is on customary units in word problems for application of fraction concepts. 4.MD.3 is addressed in Module 3.

3.NF.3 Explain equivalence of fractions in special cases, and compare fractions by reasoning about their size.

    a. Understand two fractions as equivalent (equal) if they are the same size, or the same point on a number line.

    b. Recognize and generate simple equivalent fractions, (e.g., 1/2 = 2/4, 4/6 = 2/3). Explain why the fractions are equivalent, e.g., by using a visual fraction model.

    c. Express whole numbers as fractions, and recognize fractions that are equivalent to whole numbers. *Examples: Express 3 in the form 3 = 3/1; recognize that 6/1 = 6; locate 4/4 and 1 at the same point of a number line diagram.*

3.MD.2 Measure and estimate liquid volumes and masses of objects using standard units of grams (g), kilograms (kg), and liters (l). Add, subtract, multiply, or divide to solve one-step word problems involving masses or volumes that are given in the same units, e.g., by using drawings (such as a beaker with a measurement scale) to represent the problem.

Focus Standards for Mathematical Practice

MP.2 **Reason abstractly and quantitatively.** Students create conversion charts for related measurement units and use the information in the charts to solve complex real-world measurement problems. They also draw number lines and tape diagrams to represent word problems.

MP.3 **Construct viable arguments and critique the reasoning of others.** Students work in groups to select appropriate strategies to solve problems. They present these strategies to the class and discuss the advantages and disadvantages of each strategy in different situations before deciding which ones are most efficient for that specific situation. Students also solve problems created by classmates and explain to the problem's creator how they solved it to see if it is the method the student had in mind when writing the problem.

MP.7 **Look for and make use of structure.** Students look for and make use of connections between measurement units and word problems to help them understand and solve related word problems. They choose the appropriate unit of measure when given the choice and see that the structure of the situations in the word problems dictates which units to measure with.

MP.8 **Look for and express regularity in repeated reasoning.** The creation and use of the measurement conversion tables is a focal point of this module. Students identify and use the patterns found in each table they create. Using the tables to solve various word problems gives students ample opportunities to apply the same strategy to different situations.

Overview of Module Topics and Lesson Objectives

Standards		Topics and Objectives	Days
4.OA.1 4.OA.2 4.MD.1 4.NBT.5 4.MD.2	A	**Measurement Conversion Tables** Lessons 1–2: Create conversion tables for length, weight, and capacity units using measurement tools, and use the tables to solve problems. Lesson 3: Create conversion tables for units of time, and use the tables to solve problems. Lesson 4: Solve multiplicative comparison word problems using measurement conversion tables. Lesson 5: Share and critique peer strategies.	5
4.OA.2 4.OA.3 4.MD.1 4.MD.2 4.NBT.5 4.NBT.6	B	**Problem Solving with Measurement** Lesson 6: Solve problems involving mixed units of capacity. Lesson 7: Solve problems involving mixed units of length. Lesson 8: Solve problems involving mixed units of weight. Lesson 9: Solve problems involving mixed units of time. Lessons 10–11: Solve multi-step measurement word problems.	6
4.OA.3 4.MD.1 4.MD.2 4.NBT.5 4.NBT.6	C	**Investigation of Measurements Expressed as Mixed Numbers** Lessons 12–13: Use measurement tools to convert mixed number measurements to smaller units. Lesson 14: Solve multi-step word problems involving converting mixed number measurements to a single unit.	3
		End-of-Module Assessment: Topics A–C (assessment 1 day, ½ day return, remediation or further application ½ day)	2
	D	**Year in Review** Lessons 15–16: Create and determine the area of composite figures. Lesson 17: Practice and solidify Grade 4 fluency. Lesson 18: Practice and solidify Grade 4 vocabulary.	4
Total Number of Instructional Days			**20**

Module 7: Exploring Measurement with Multiplication

Terminology

New or Recently Introduced Terms

- Cup (c) (customary unit of measure for liquid volume)
- Customary system of measurement (measurement system commonly used in the United States that includes such units as yards, pounds, and gallons)
- Customary unit (e.g., foot, ounce, quart)
- Gallon (gal) (customary unit of measure for liquid volume)
- Metric system of measurement (base-ten system of measurement used internationally that includes such units as meters, kilograms, and liters)
- Metric unit (e.g., kilometer, gram, milliliter)
- Ounce (oz) (customary unit of measure for weight)
- Pint (pt) (customary unit of measure for liquid volume)
- Pound (lb) (customary unit of measure for weight)
- Quart (qt) (customary unit of measure for liquid volume)

Familiar Terms and Symbols[2]

- Capacity (the maximum amount that a container can hold)
- Convert (to express a measurement in a different unit)
- Distance (the length of the line segment joining two points)
- Equivalent (the same)
- Foot (ft) (customary unit of measure for length)
- Gram (g), kilogram (kg) (metric units of measure for mass, not distinguished from weight at this time)
- Hour (hr) (unit of measure for time)
- Inch (customary unit of measure for length, 12 inches = 1 foot)
- Interval (time passed or a segment on the number line)
- Length (the measurement of something from end to end)
- Liter (L), milliliter (mL) (metric units of measure for liquid volume)
- Measurement (dimensions, quantity, or capacity as determined by comparison with a standard)
- Meter (m), centimeter (cm), kilometer (km) (metric units of measure for length)
- Minute (min) (unit of measure for time)
- Mixed units (e.g., 3 m 43 cm)
- Second (sec) (unit of measure for time)
- Table (used to represent data)
- Weight (the measurement of how heavy something is)
- Yard (yd) (customary unit of measure for length)

---

[2]These are terms and symbols students have seen previously.

Suggested Tools and Representations

- Analog clock (with second hand)
- Balance scale with mass weights
- Beaker (marked for mL and L)
- Composite figure
- Digital scale (metric and customary units)
- Gallon, quart, pint, and cup containers
- Meter stick, yard stick, 12-inch ruler, centimeter ruler
- Number bond
- Number line
- Protractor
- Stopwatch
- Tape diagram
- Two-column table

Scaffolds[3]

The scaffolds integrated into *A Story of Units* give alternatives for how students access information as well as express and demonstrate their learning. Strategically placed margin notes are provided within each lesson elaborating on the use of specific scaffolds at applicable times. They address many needs presented by English language learners, students with disabilities, students performing above grade level, and students performing below grade level. Many of the suggestions are organized by Universal Design for Learning (UDL) principles and are applicable to more than one population. To read more about the approach to differentiated instruction in *A Story of Units*, please refer to "How to Implement *A Story of Units*."

Assessment Summary

Type	Administered	Format	Standards Addressed
End-of-Module Assessment Task	After Topic C	Constructed response with rubric	4.OA.1 4.OA.2 4.OA.3 4.MD.1 4.MD.2

---

[3]Students with disabilities may require Braille, large print, audio, or special digital files. Please visit the website www.p12.nysed.gov/specialed/aim for specific information on how to obtain student materials that satisfy the National Instructional Materials Accessibility Standard (NIMAS) format.

Module 7: Exploring Measurement with Multiplication

A STORY OF UNITS

Mathematics Curriculum

GRADE 4 • MODULE 7

Topic A
Measurement Conversion Tables

4.OA.1, 4.OA.2, 4.MD.1, 4.NBT.5, 4.MD.2

Focus Standards:	4.OA.1	Interpret a multiplication equation as a comparison, e.g., interpret 35 = 5 × 7 as a statement that 35 is 5 times as many as 7 and 7 times as many as 5. Represent verbal statements of multiplicative comparisons as multiplication equations.
	4.OA.2	Multiply or divide to solve word problems involving multiplicative comparison, e.g., by using drawings and equations with a symbol for the unknown number to represent the problem, distinguishing multiplicative comparison from additive comparison. (See CCSS-M Glossary, Table 2.)
	4.MD.1	Know relative sizes of measurement units within one system of units including km, m, cm; kg, g; lb, oz.; l, ml; hr, min, sec. Within a single system of measurement, express measurements in a larger unit in terms of a smaller unit. Record measurement equivalents in a two-column table. *For example, know that 1 ft is 12 times as long as 1 in. Express the length of a 4 ft snake as 48 in. Generate a conversion table for feet and inches listing the number pairs (1, 12), (2, 24), (3, 36), …*
Instructional Days:	5	
Coherence -Links from:	G3–M1	Properties of Multiplication and Division and Solving Problems with Units of 2–5 and 10
	G3–M2	Place Value and Problem Solving with Units of Measure
-Links to:	G5–M1	Place Value and Decimal Fractions
	G5–M2	Multi-Digit Whole Number and Decimal Fraction Operations

In Topic A, students build on the work they did in Module 2 with measurement conversions. In this module, however, they have the opportunity to work more extensively with tools while creating two-column tables that are then used to solve a variety of measurement problems.

In Lesson 1, students use two-column conversion tables (**4.MD.1**) to practice conversion rates. Students convert from pounds to ounces, yards to feet, and feet to inches. Students begin Lesson 1 by using a balance scale, a 1-pound weight, and individual 1-ounce weights (like fishing sinkers). With the 1-pound weight on one side of the balance, they add 1 ounce at a time to the other side until it balances to discover that there are 16 ounces in 1 pound. Students then generate a two-column conversion table listing the number of ounces in 2, 3, and up to 10 pounds. Students use their multiplication skills from Module 3 to complete the table and reason about why they do not need to complete the table beyond 10 pounds.

Students use various strategies to determine how many smaller units would make up a larger unit not listed in the table. A student could reason, for example, that since the table shows that there are 160 ounces in 10 pounds and 48 ounces in 3 pounds, he can add them together to tell that there are 208 ounces in 13 pounds. Another student might reason that since there are 16 ounces in each pound, she can use the rule of the table and multiply 13 pounds by 16 to find that there are 208 ounces in 13 pounds.

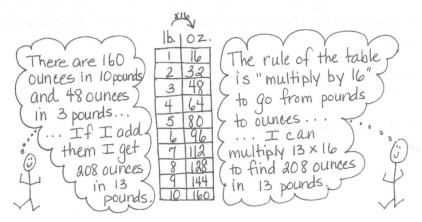

Similar to Lesson 1, in Lesson 2 students complete conversion tables, this time focusing on capacity and converting gallons to quarts, quarts to pints, and pints to cups. Adding to the complexity of the conversions, students explore two-step conversions, solving, for example, to find how many cups are equal to 1 gallon.

In Lesson 3, students investigate the relationships between units of time. They discover a similarity in converting from hours to minutes and minutes to seconds. Students are able to reason that, for both sets of conversions, the values in the two tables will be the same because there are 60 seconds in a minute and 60 minutes in an hour. Students also convert from days to hours. The clock and the number line are used as tools to develop the conversion tables.

In Lesson 4, students use the conversions that they discovered in Lessons 1–3 to solve multiplicative comparison word problems. Working in small groups, students have the opportunity to share and discuss their solution strategies (**4.OA.1, 4.OA.2**).

After being given tape diagrams, students are challenged to create word problems to match the information displayed within the tape diagrams in Lesson 5. The given information requires students to use the customary or metric units practiced during this topic. After first solving the problems that they create, students share and critique the problem-solving strategies used by their peers. In the Debrief, by way of answering guided questions, students discuss not only how the problems were solved but also the advantages and disadvantages of using each strategy. They further discuss what makes one strategy more efficient than another.

A STORY OF UNITS   Topic A   4•7

A Teaching Sequence Toward Mastery of Measurement Conversion Tables

Objective 1: Create conversion tables for length, weight, and capacity units using measurement tools, and use the tables to solve problems.
(Lessons 1–2)

Objective 2: Create conversion tables for units of time, and use the tables to solve problems.
(Lesson 3)

Objective 3: Solve multiplicative comparison word problems using measurement conversion tables.
(Lesson 4)

Objective 4: Share and critique peer strategies.
(Lesson 5)

Lesson 1

Objective: Create conversion tables for length, weight, and capacity units using measurement tools, and use the tables to solve problems.

Suggested Lesson Structure

- Fluency Practice   (12 minutes)
- Concept Development (38 minutes)
- Student Debrief   (10 minutes)
- **Total Time**   **(60 minutes)**

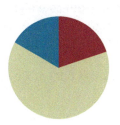

Fluency Practice (12 minutes)

- Sprint: Convert to Dollars **4.MD.2**   (9 minutes)
- Add and Subtract **4.NBT.4**   (3 minutes)

Sprint: Money (9 minutes)

Materials: (S) Convert to Dollars Sprint

Note: This Sprint reviews Module 6 Topic E.

Add and Subtract (3 minutes)

Materials: (S) Personal white board

Note: This fluency activity reviews adding and subtracting using the standard algorithm.

- T: (Write 699 thousands 999 ones.) On your personal white board, write this number in standard form.
- S: (Write 699,999.)
- T: (Write 155 thousands 755 ones.) Add this number to 699,999 using the standard algorithm.
- S: (Solve 699,999 + 155,755 = 855,754 using the standard algorithm.)

Continue the process for 456,789 + 498,765.

- T: (Write 400 thousand 1 one.) On your board, write this number in standard form.
- S: (Write 400,001.)

> **NOTES ON MULTIPLE MEANS OF ACTION AND EXPRESSION:**
>
> Challenge students working above grade level and others to apply efficient alternative strategies learned since Grade 1 to solve the Add and Subtract fluency activity.

A STORY OF UNITS   Lesson 1 4•7

T: (Write 235 thousands 165 ones.) Subtract this number from 400,001 using the standard algorithm.
S: (Solve 400,001 − 235,165 = 164,836 using the standard algorithm.)

Continue the process for 708,050 − 256,089.

Concept Development (38 minutes)

Materials: (T) Balance scale, 1-pound weights, 1-ounce weights, yardstick, ruler (S) Balance scale (1 per group), 1-pound weight (1 per group), 1-ounce weights (16 per group), yardstick (1 per group), ruler (1 per group), Practice Sheet

Note: Groups of 3 students are suggested for this Concept Development.

Problem 1: Convert pounds to ounces.

Display the words *pound* and *ounce*.

T: (Hold up one 1-ounce weight.) This item weighs 1 **ounce**. I am going to place it on one side of the scale. (Place it on the scale.)
T: (Hold up one 1-pound weight.) This item weighs 1 **pound**. I am going to place it on the other side of the scale. (Place it on the scale.) What do you notice?
S: The scale moved! → The pound must weigh more than the ounce because the scale went down on the pound's side when you added that weight.
T: In your groups, use the scale and weights to determine how many ounces equal 1 pound.

Allow time for every group to reach a conclusion.

T: How many 1-ounce weights did you need to balance the scale?
S: 16 1-ounce weights. → 16 ounces.

T: (Display the two-column table.) Now, we know that 1 pound equals 16 ounces. Fill in the first line of the table.
T: How can we determine how many ounces are in 2 pounds?
S: We can add another 1-pound weight and see how many more 1-ounce weights we need to balance. → We can just double 16 ounces or multiply by 2. → 16 ounces times 2 is 32 ounces.
T: Fill in the rest of the conversion table for converting pounds to ounces. (Allow students time to work.)
T: Looking at the table, what is the rule for converting pounds to ounces?
S: Keep adding 16. → Take the number of pounds and multiply it by 16. → 1 pound is 16 ounces.
T: How can we determine how many ounces are in 15 pounds?
S: We can make the table longer and go all the way to 15 pounds. → We can multiply 15 pounds by 16. → We can add the number of ounces in 10 pounds and 5 pounds together! → We can multiply the number of ounces in 5 pounds by 3.

Pounds	Ounces
1	16
2	32
3	48
4	64
5	80
6	96
7	112
8	128
9	144
10	160

Lesson 1: Create conversion tables for length, weight, and capacity units using measurement tools, and use the tables to solve problems.

A STORY OF UNITS   Lesson 1   4•7

T: Take a moment to calculate the number of ounces in 15 pounds.

S: 15 pounds is equal to 240 ounces.

T: Convert 12 pounds 10 ounces into ounces. Discuss with your partner while working.

MP.8

S: First, we need to convert the pounds to ounces and then add 10 ounces more. → Let's do 12 × 16 ounces and add 10 ounces. → We can use the conversion table to add the ounces in 10 pounds and 2 pounds. Then, we can add 10 ounces. 192 ounces + 10 ounces = 202 ounces.

T: Pounds and ounces are part of a system called the **customary system of measurement**. In the United States, we've historically used **customary units**, such as pounds and ounces. In other countries, and more and more often now in the United States, the **metric system of measurement** is used. We've studied and have used **metric units** this year when we've solved word problems and converted kilometers to meters, kilograms to grams, and liters to milliliters.

> **NOTES ON MULTIPLE MEANS OF ENGAGEMENT:**
>
> Using alternative strategies to solve for ounces can engage students working below grade level. Assigning mixed-ability groups is also an option, but watch for stronger students assuming the role of calculating every product or sum. One strategy may be to solve for doubles. For example, if students find the number of ounces for 4 pounds, they can double that number to solve for 8 pounds, and so on. Give successful students working above grade level and others an opportunity to share their efficient strategies.

Problem 2: Convert yards to feet.

T: In your groups, compare the yardstick to the foot ruler and share what you notice.

S: A yardstick is 36 inches, but the ruler is only 12 inches. → It takes 3 rulers to be equal to 1 yardstick. → 1 yard is 3 times as long as 1 foot.

T: (Display the two-column table.) On your Practice Sheet, fill in 1 yard equals 3 feet.

T: How many feet are in 2 yards?

S: 6 feet!

T: Complete the table. (Allow students time to work.)

T: Now, find the number of feet in 37 yards 2 feet. Work until you have found the answer, and then share your strategy.

S: I multiplied 37 by 3 and added 2 feet. 111 feet + 2 feet is 113 feet. → 37 yards = 30 yards + 7 yards = (30 × 3) feet + 21 feet = 111 feet. 111 feet + 2 feet = 113 feet. → I used the distributive property. 37 yards = 30 yards + 7 yards = (30 × 3) feet + (7 × 3) feet = 90 feet + 21 feet = 111 feet. 111 feet + 2 feet = 113 feet.

Yards	Feet
1	3
2	6
3	9
4	12
5	15
6	18
7	21
8	24
9	27
10	30

Problem 3: Convert feet to inches.

T: In your groups, examine the ruler, and share what you notice about the relationship between inches and feet.

Feet	Inches
1	12
2	24
3	36
4	48
5	60
6	72
7	84
8	96
9	108
10	120

Lesson 1: Create conversion tables for length, weight, and capacity units using measurement tools, and use the tables to solve problems.

A STORY OF UNITS   Lesson 1   4•7

S: Inches are smaller than feet. → There are 12 inches in 1 foot.

T: We know that 1 foot equals 12 inches. On the table, fill in the first line.

T: Continue to fill out the table just like we did for the other units. (Allow students time to work.)

T: 1 foot is how many times the length of an inch?

S: 12 times.

T: Talk to your partner. How could you find out how many inches are in 20 feet?

S: We know that 10 feet equals 120 inches. We can just double 120 to get 240 inches in 20 feet. → We could multiply 20 times 12 because there are 12 inches in 1 foot.

Follow up by having students find the number of inches in 6 feet 8 inches, 25 feet 5 inches, and 32 feet 7 inches.

Problem Set (10 minutes)

Students should do their personal best to complete the Problem Set within the allotted 10 minutes. Some problems do not specify a method for solving. This is an intentional reduction of scaffolding that invokes MP.5, Use Appropriate Tools Strategically. Students should solve these problems using the RDW approach used for Application Problems.

For some classes, it may be appropriate to modify the assignment by specifying which problems students should work on first. With this option, let the careful sequencing of the Problem Set guide the selections so that problems continue to be scaffolded. Balance word problems with other problem types to ensure a range of practice. Assign incomplete problems for homework or at another time during the day.

Student Debrief (10 minutes)

Lesson Objective: Create conversion tables for length, weight, and capacity units using measurement tools, and use the tables to solve problems.

The Student Debrief is intended to invite reflection and active processing of the total lesson experience.

Invite students to review their solutions for the Problem Set. They should check work by comparing answers with a partner before going over answers as a class. Look for misconceptions or misunderstandings that can be addressed in the Debrief. Guide students in a conversation to debrief the Problem Set and process the lesson.

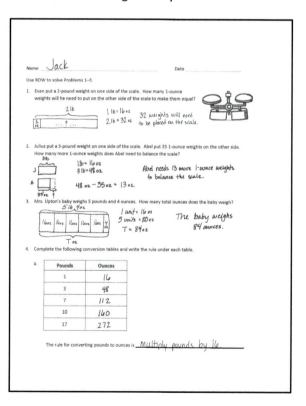

Any combination of the questions below may be used to lead the discussion.

- What strategy did you use to solve Problem 2? Did you need the conversion table to help you convert **pounds** to **ounces**? If not, what rule did you use?

16   Lesson 1:   Create conversion tables for length, weight, and capacity units using measurement tools, and use the tables to solve problems.

A STORY OF UNITS

Lesson 1 4•7

- Explain your solution for Problem 5(h) to your partner. Is there a rule for converting yards to inches?
- When might you need to compare units in real life like those in Problem 6?
- Looking at the conversion tables, what do you notice about the units that we are converting?
- Is it easier to use the conversion table or to use the rule to convert? Why?
- Name some units that are **customary units**. Name some units that are **metric units**.
- A yard and a meter are close in length but not exactly the same. Yards are part of the **customary system of measurement,** and meters are part of the **metric system of measurement**. Can you think of any other pairs that are close, but not the same, like this?

Exit Ticket (3 minutes)

After the Student Debrief, instruct students to complete the Exit Ticket. A review of their work will help with assessing students' understanding of the concepts that were presented in today's lesson and planning more effectively for future lessons. The questions may be read aloud to the students.

b.

Feet	Inches
1	12
2	24
5	60
10	120
15	180

The rule for converting feet to inches is
multiply feet times 12.

c.

Yards	Feet
1	3
2	6
4	12
10	30
14	42

The rule for converting yards to feet is
multiply yards times 3.

5. Solve.
   a. 3 feet 1 inch = __37__ inches
   b. 11 feet 10 inches = __142__ inches
   c. 5 yards 1 foot = __16__ feet
   d. 12 yards 2 feet = __38__ feet
   e. 27 pounds 10 ounces = __442__ ounces
   f. 18 yards 9 feet = __63__ feet
   g. 14 pounds 5 ounces = __229__ ounces
   h. 5 yards 2 feet = __204__ inches

6. Answer "true" or "false" for the following statements. If the statement is false, change the right side of the comparison to make it true.
   a. 2 kilograms > 2,600 grams __false__
      2 kilograms > 1,600 grams
   b. 12 feet < 140 inches __false__
      12 feet < 150 inches
   c. 10 kilometers = 10,000 meters __true__

Lesson 1: Create conversion tables for length, weight, and capacity units using measurement tools, and use the tables to solve problems.

A STORY OF UNITS Lesson 1 Sprint 4•7

A

Number Correct: _____

Convert to Dollars

1.	1 cent =	$ 0.		23.	6 pennies =	
2.	2 cents =			24.	5 dimes =	
3.	3 cents =			25.	5 pennies =	
4.	8 cents =			26.	1 dime 1 penny =	
5.	80 cents =			27.	1 dime 2 pennies =	
6.	70 cents =			28.	1 dime 7 pennies =	
7.	60 cents =			29.	4 dimes 5 pennies =	
8.	20 cents =			30.	6 dimes 3 pennies =	
9.	1 penny =			31.	3 pennies 6 dimes =	
10.	1 dime =			32.	7 pennies 9 dimes =	
11.	2 pennies =			33.	1 quarter =	
12.	2 dimes =			34.	2 quarters =	
13.	3 pennies =			35.	3 quarters =	
14.	3 dimes =			36.	2 quarters 3 pennies =	
15.	9 dimes =			37.	1 quarter 3 pennies =	
16.	7 pennies =			38.	3 quarters 3 pennies =	
17.	8 dimes =			39.	2 quarters 2 dimes =	
18.	4 pennies =			40.	1 quarter 1 dime =	
19.	6 dimes =			41.	3 quarters 1 dime =	
20.	8 pennies =			42.	1 quarter 4 dimes =	
21.	7 dimes =			43.	3 quarters 2 dimes =	
22.	9 pennies =			44.	3 quarters 18 pennies =	

Lesson 1: Create conversion tables for length, weight, and capacity units using measurement tools, and use the tables to solve problems.

EUREKA MATH

B

Convert to Dollars

Number Correct: _____

Improvement: _____

#	Problem	Answer	#	Problem	Answer
1.	2 cents =	$ 0.	23.	5 pennies =	
2.	3 cents =		24.	6 dimes =	
3.	4 cents =		25.	4 pennies =	
4.	9 cents =		26.	1 dime 1 penny =	
5.	90 cents =		27.	1 dime 2 pennies =	
6.	80 cents =		28.	1 dime 8 pennies =	
7.	70 cents =		29.	5 dimes 4 pennies =	
8.	30 cents =		30.	7 dimes 4 pennies =	
9.	1 penny =		31.	4 pennies 7 dimes =	
10.	1 dime =		32.	6 pennies 8 dimes =	
11.	2 pennies =		33.	1 quarter =	
12.	2 dimes =		34.	2 quarters =	
13.	3 pennies =		35.	3 quarters =	
14.	3 dimes =		36.	2 quarters 4 pennies =	
15.	8 dimes =		37.	1 quarter 4 pennies =	
16.	6 pennies =		38.	3 quarters 4 pennies =	
17.	7 dimes =		39.	2 quarters 3 dimes =	
18.	9 pennies =		40.	1 quarter 2 dimes =	
19.	5 dimes =		41.	3 quarters 2 dimes =	
20.	7 pennies =		42.	1 quarter 5 dimes =	
21.	9 dimes =		43.	3 quarters 1 dime =	
22.	8 pennies =		44.	3 quarters 19 pennies =	

Name _____   Date _____

a.

Pounds	Ounces
1	
2	
3	
4	
5	
6	
7	
8	
9	
10	

The rule for converting pounds to ounces is _____.

b.

Yards	Feet
1	
2	
3	
4	
5	
6	
7	
8	
9	
10	

The rule for converting yards to feet is _____.

c.

Feet	Inches
1	
2	
3	
4	
5	
6	
7	
8	
9	
10	

The rule for converting feet to inches is _____.

Lesson 1: Create conversion tables for length, weight, and capacity units using measurement tools, and use the tables to solve problems.

Name _____ Date _____

Use RDW to solve Problems 1–3.

1. Evan put a 2-pound weight on one side of the scale. How many 1-ounce weights will he need to put on the other side of the scale to make them equal?

2. Julius put a 3-pound weight on one side of the scale. Abel put 35 1-ounce weights on the other side. How many more 1-ounce weights does Abel need to balance the scale?

3. Mrs. Upton's baby weighs 5 pounds and 4 ounces. How many total ounces does the baby weigh?

4. Complete the following conversion tables, and write the rule under each table.

   a.

Pounds	Ounces
1	
3	
7	
10	
17	

   The rule for converting pounds to ounces is _____.

b.

Feet	Inches
1	
2	
5	
10	
15	

The rule for converting feet to inches is

_____.

c.

Yards	Feet
1	
2	
4	
10	
14	

The rule for converting yards to feet is

_____.

5. Solve.

   a. 3 feet 1 inch = _____ inches

   b. 11 feet 10 inches = _____ inches

   c. 5 yards 1 foot = _____ feet

   d. 12 yards 2 feet = _____ feet

   e. 27 pounds 10 ounces = _____ ounces

   f. 18 yards 9 feet = _____ feet

   g. 14 pounds 5 ounces = _____ ounces

   h. 5 yards 2 feet = _____ inches

6. Answer *true* or *false* for the following statements. If the statement is false, change the right side of the comparison to make it true.

   a. 2 kilograms > 2,600 grams   _____

   b. 12 feet < 140 inches   _____

   c. 10 kilometers = 10,000 meters   _____

Name _____ Date _____

1. Solve.

   a. 8 feet = _____ inches

   b. 4 yards 2 feet = _____ feet

   c. 14 pounds 7 ounces = _____ ounces

2. Answer *true* or *false* for the following statements. If the statement is false, change the right side of the comparison to make it true.

   a. 3 pounds > 60 ounces   _____

   b. 12 yards < 40 feet   _____

Name _____   Date _____

1. Complete the tables.

   a.

Yards	Feet
1	
2	
3	
5	
10	

   b.

Feet	Inches
1	
2	
5	
10	
15	

   c.

Yards	Inches
1	
3	
6	
10	
12	

2. Solve.

   a. 2 yards 2 inches = _____ inches

   b. 9 yards 10 inches = _____ inches

   c. 4 yards 2 feet = _____ feet

   d. 13 yards 1 foot = _____ feet

   e. 17 feet 2 inches = _____ inches

   f. 11 yards 1 foot = _____ feet

   g. 15 yards 2 feet = _____ feet

   h. 5 yards 2 feet = _____ inches

3. Ally has a piece of string that is 6 yards 2 feet long. How many inches of string does she have?

4. Complete the table.

Pounds	Ounces
1	
2	
4	
10	
12	

5. Renee's baby sister weighs 7 pounds 2 ounces. How many ounces does her sister weigh?

6. Answer *true* or *false* for the following statements. If the statement is false, change the right side of the comparison to make it true.

   a. 4 kilograms < 4,100 grams   _____

   b. 10 yards < 360 inches   _____

   c. 10 liters = 100,000 milliliters   _____

Lesson 2

Objective: Create conversion tables for length, weight, and capacity units using measurement tools, and use the tables to solve problems.

Suggested Lesson Structure

■ Fluency Practice   (11 minutes)
■ Concept Development (39 minutes)
■ Student Debrief   (10 minutes)
Total Time   **(60 minutes)**

Fluency Practice (11 minutes)

- Grade 4 Core Fluency Differentiated Practice Sets **4.NBT.4**   (5 minutes)
- Convert Length Units **4.MD.1**   (4 minutes)
- Convert Capacity Units **4.MD.1**   (2 minutes)

Grade 4 Core Fluency Differentiated Practice Sets (5 minutes)

Materials: (S) Core Fluency Practice Sets

Note: In this lesson and throughout Module 7, Fluency Practice includes an opportunity for review and mastery of the addition and subtraction algorithm by means of the Core Fluency Practice Sets. Four options are provided in this lesson:

- Practice Set A is multi-digit addition.
- Practice Set B is multi-digit subtraction.
- Practice Set C is multi-digit subtraction with zeros in the minuend.
- Practice Set D is multi-digit addition and subtraction.

All Practice Sets have a Part 1 and a Part 2. Note that Part 2 has fewer regroupings and may be used for students working below grade level. The answers to both Part 1 and Part 2 are the same for ease of correction.

Students complete as many problems as possible in 120 seconds. Collect any Practice Sets that have been completed within the 120 seconds, and check the answers. Students who do not finish in 120 seconds can be encouraged to use their Practice Sets for practice at home or for remedial practice in the classroom. The next time the Practice Sets are used, students who have successfully completed their set with 100% accuracy can move to the next level. Others should repeat the same level until mastery. Keep a record of student progress.

For early finishers, assign a counting pattern and start number. For example, "Finish early? Count by sevens starting at 168 on the back of your Practice Set." Celebrate improvement and advancement. Encourage students to compete with themselves rather than with their peers. Notify caring adults of each child's progress.

Convert Length Units (4 minutes)

Materials: (S) Personal white board

Note: This fluency activity reviews Lesson 1 and metric conversions from Module 2.

    T: (Write 1 km = ___ m.) How many meters are in 1 kilometer?
    S: 1,000 m.

Repeat the process for 2 and 3 kilometers.

    T: (Write 7 km = ___ m.) Write the number sentence.
    S: (Write 7 km = 7,000 m.)
    T: (Write 1 m = ___ cm.) How many centimeters are in 1 meter?
    S: 100 cm.

Repeat the process for 2 and 3 meters.

    T: (Write 8 m = ___ cm.) Write the number sentence.
    S: (Write 8 m = 800 cm.)
    T: (Write 1 yd = ___ ft.) How many feet are in 1 yard?
    S: 3 feet.

Repeat the process for 2 and 3 yards.

    T: (Write 10 yd = ___ ft.) Write the number sentence.
    S: (Write 10 yd = 30 ft.)
    T: (Write 1 ft = ___ in.) How many inches are in 1 foot?
    S: 12 inches.

Repeat the process for 2 and 3 feet.

Convert Capacity Units (2 minutes)

Materials: (S) Personal white board

Note: This fluency activity reviews metric conversions from Module 2.

    T: (Write 1 liter = ___ mL.) How many milliliters are in 1 liter?
    S: 1,000 mL.

Repeat the process for 2 and 3 liters.

    T: (Write 6 liters = ___ mL.) Write the number sentence.
    S: (Write 6 liters = 6,000 mL.)

A STORY OF UNITS   Lesson 2   4•7

Concept Development (39 minutes)

Materials: (T) Gallon container, quart container, pint container, liquid measuring cup, funnel, water
(S) Gallon container (1 per group), quart container (1 per group), pint container (1 per group), liquid measuring cup (1 per group), funnel (1 per group), bucket filled with 1.5 gallons of water (1 per group), Practice Sheet

Note: Groups of 3 are recommended.

Problem 1: Convert gallons to quarts.

T: (Hold up a 1-gallon container and a 1-quart container. Display the words **gallon** and **quart**.) This container measures 1 gallon. This container measures 1 quart of liquid. In your groups, fill the gallon with water using a 1-quart measure. How many quarts does it take to fill a gallon?

Allow enough time for all groups to have an opportunity to fill a gallon.

S: We used 4 quarts to fill the gallon.

T: On your Practice Sheet, fill in the first line of the Gallons to Quarts table. What is the rule for converting gallons to quarts?

S: Multiply the number of gallons by 4. → A gallon holds 4 times as much as a quart.

T: How can we find the number of quarts in 13 gallons?

S: We could get 13 1-gallon containers and fill them up using the quart measure. → We can use the rule and multiply the number of gallons by 4. 13 × 4 = 52. 13 gallons = 52 quarts. → The conversion table tells us how many quarts are in 3 gallons and 10 gallons. Just add them together.

T: Just like we converted pounds to ounces, yards to feet, and feet to inches in the last lesson, we can use the conversion table as a tool to solve for the number of gallons in a given number of quarts.

T: How many quarts will there be in 18 gallons 3 quarts?

S: 75 quarts.

Gallons	Quarts
1	4
2	8
3	12
4	16
5	20
6	24
7	28
8	32
9	36
10	40

Problem 2: Convert quarts to pints. Relate pints to gallons.

T: (Hold up a 1-pint measure, and display the word **pint**.) This is a measurement of 1 pint. Fill the quart to find out how many pints equal 1 quart.

Allow enough time for all groups to have an opportunity to fill a quart.

S: We used 2 pints to fill the quart.

T: (Display the two-column table.) We now know that 1 quart equals 2 pints. On your Practice Sheet, fill in the first line.

T: What is the rule for converting quarts to pints?

Quarts	Pints
1	2
2	4
3	6
4	8
5	10
6	12
7	14
8	16
9	18
10	20

A STORY OF UNITS　　　　　　　　　　　　　　　　　　　　　　　　　　　　Lesson 2　4•7

S: Multiply the number of quarts by 2. → A quart holds 2 times as much as a pint.
T: Complete the table. Then, solve for the number of pints in 13 quarts 1 pint.
T: How did you find the answer?
S: There are 26 pints in 13 quarts, plus 1 pint is 27 pints. → There are 20 pints in 10 quarts and 6 pints in 3 quarts, so 20 pints plus 6 pints is 26 pints. Last, we have to add the 1 pint to get 27 pints in 13 quarts 1 pint.
T: (Point to the row with 4 quarts.) What do we know about the relationship between quarts and gallons?
S: 4 quarts equal 1 gallon.
T: So how many pints equal 1 gallon?
S: 8 pints equal 1 gallon.
T: Pour 8 pints into 1 gallon to confirm.

Allow time for students to test their answers.

T: With your partner, solve for the number of pints in 2 gallons 3 quarts.
S: 2 gallons is 16 pints. 3 quarts is 6 pints. 22 pints equal 2 gallons 3 quarts.

Problem 3: Convert pints to cups. Relate cups to quarts and gallons.

T: (Hold up a 1-cup measure, and display the word **cup**.) This is a measurement of 1 cup. Fill the pint to find out how many cups equal 1 pint.

Allow enough time for all groups to have an opportunity to fill a pint. Repeat as done in Problems 1 and 2 by finding the conversion rule, completing the table on the Practice Sheet, and solving for the number of cups in 16 pints 1 cup.

T: (Point to the row with 2 pints.) What do we know about the relationship between pints and quarts?
S: 2 pints equal 1 quart.
T: So, how many cups equal 1 quart?
S: 4 cups equal 1 quart.
T: Pour 4 cups into 1 quart to confirm. Discuss the conversion rule for quarts to cups with your partner. (Allow time for students to test their answers.)
S: Multiply the number of quarts by 4. → A quart is 4 times as much as a cup.
T: With your partner, solve for the number of cups in 3 quarts 1 pint.
S: 1 quart is 4 cups, so 3 quarts is 12 cups. 1 pint is 2 cups. 14 cups equal 3 quarts 1 pint.

> **NOTES ON MULTIPLE MEANS OF ACTION AND EXPRESSION:**
>
> Some learners may need scaffolds for managing information, such as conversion rules for measurement units. Provide graphic organizers and other visuals, such as the four-column table below, to assist students with connecting relationships between gallons, quarts, pints, and cups.
>
> | Gallons | Quarts | Pints | Cups |
> |---------|--------|-------|------|
> | | | 1 | 2 |
> | | 1 | 2 | 4 |
> | | | 3 | 6 |
> | 1 | 4 | 8 | 16 |

Pints	Cups
1	2
2	4
3	6
4	8
5	10
6	12
7	14
8	16
9	18
10	20

Lesson 2: Create conversion tables for length, weight, and capacity units using measurement tools, and use the tables to solve problems.

29

T: With your partner, use the conversion rules to determine how many cups equal 1 gallon.

S: 1 quart equals 4 cups. 4 quarts equal 1 gallon. So, 16 cups equal 1 gallon.

T: Measure to confirm. Complete the statements comparing cups, pints, quarts, and gallons on the Practice Sheet.

Display: 1 gallon = 4 quarts, 1 quart = 4 cups, 1 quart = 2 pints, 1 pint = 2 cups.

T: Discuss with your partner some of these rules we learned today. How will you be able to keep them organized? What relationships do you see between these measurements?

S: Cups, quarts, and gallons are closely related because it takes 4 cups to equal 1 quart and 4 quarts to make 1 gallon. So, the rule of *times 4* makes those measurements easy to memorize and use.
→ Quarts, pints, and cups are closely related, too. 2 cups equal 1 pint. 2 pints equal 1 quart. If I use the *times 2* rule with these measurements, they will be easy to convert.

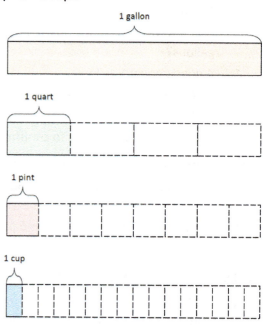

Support students in understanding the connectedness of these 4 capacity measurements by displaying an image, such as the one pictured to the right, or by asking students to draw tape diagrams to show the relationship if they are unable to express it in words.

Problem 4: Calculate a two-step conversion.

Brandon made 5 gallons of iced tea for his party. How many cups of iced tea can he serve?

T: With your partner, use the RDW process to solve this problem.

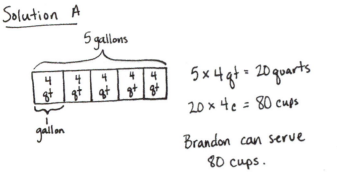

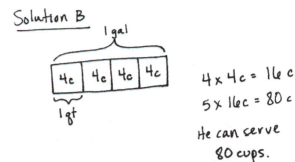

In Solution A, the student starts with the whole tape representing 5 gallons, finds the number of quarts in all the iced tea, and then multiplies the number of quarts by 4 to find the total number of cups. In Solution B, the student finds the number of cups in 1 gallon and then multiplies that by 5 to find the number of cups in 5 gallons. Be aware of students who know 1 gallon equals 16 cups and multiply 16 times 5.

Problem Set (10 minutes)

Students should do their personal best to complete the Problem Set within the allotted 10 minutes. For some classes, it may be appropriate to modify the assignment by specifying which problems they work on first. Some problems do not specify a method for solving. Students should solve these problems using the RDW approach used for Application Problems.

Student Debrief (10 minutes)

Lesson Objective: Create conversion tables for length, weight, and capacity units using measurement tools, and use the tables to solve problems.

The Student Debrief is intended to invite reflection and active processing of the total lesson experience.

Invite students to review their solutions for the Problem Set. They should check work by comparing answers with a partner before going over answers as a class. Look for misconceptions or misunderstandings that can be addressed in the Debrief. Guide students in a conversation to debrief the Problem Set and process the lesson.

Any combination of the questions below may be used to lead the discussion.

- Let's compare two different representations of Problem 2. How are these students' solutions the same? How are they different? How did each represent the comparison?
- How could you solve Problem 5(c) using the conversion table? How could you solve it without a conversion table?
- Explain the strategy you used to solve Problem 8. Compare solution strategies with your small group. Whose way of solving was the most efficient?
- If someone wanted to find the number of **pints** or **quarts** in 150 **gallons**, would it make sense to use the conversion table to solve? Why is understanding the conversion rule important?

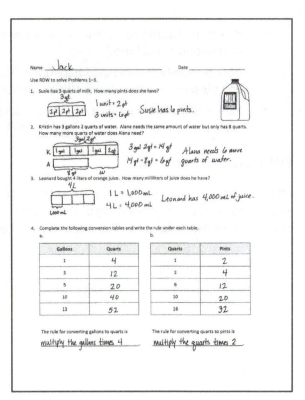

Lesson 2: Create conversion tables for length, weight, and capacity units using measurement tools, and use the tables to solve problems.

A STORY OF UNITS — Lesson 2 4•7

- When measuring liquid, such as a bottle of drinking water, which customary unit could I use? Which metric unit could I use?
- Milk is usually sold in customary units. Which customary units are used for milk? On the other hand, soda is sold in metric units. Which metric unit is used for soda?
- How many gallons of juice would our class need to serve at its next party so that each student in our class would receive 1 **cup** of juice? What about 2 cups of juice?

Exit Ticket (3 minutes)

After the Student Debrief, instruct students to complete the Exit Ticket. A review of their work will help with assessing students' understanding of the concepts that were presented in today's lesson and planning more effectively for future lessons. The questions may be read aloud to the students.

Name _____   Date _____

Practice Set A Part 1: Multi-Digit Addition Fluency

1.
```
    8,149
+   7,264
```

2.
```
   42,609
+   8,685
```

3.
```
   39,563
+  48,438
```

4.
```
  658,199
+  25,675
```

5.
```
  445,976
+  37,415
```

6.
```
  438,617
+ 493,859
```

Practice Set A Part 2: Multi-Digit Addition Fluency

1.
```
    9,202
+   6,211
```

2.
```
   42,774
+   8,520
```

3.
```
   53,545
+  34,456
```

4.
```
  604,754
+  79,120
```

5.
```
   54,315
+  29,076
```

6.
```
  110,728
+ 821,748
```

A STORY OF UNITS

Lesson 2 Core Fluency Practice Set B 4•7

Name _____ Date _____

Practice Set B Part 1: Multi-Digit Subtraction Fluency

1.
```
    7, 7 3 9
  − 5, 5 4 6
```

2.
```
   2 3, 1 4 5
  −    5, 1 2 9
```

3.
```
   7 1, 3 7 8
  − 6 1, 8 7 6
```

4.
```
   4 7 9, 5 4 1
  −   7 8, 8 5 6
```

Practice Set B Part 2: Multi-Digit Subtraction Fluency

1.
```
    7, 6 9 9
  − 5, 5 0 6
```

2.
```
   1 9, 1 4 5
  −    1, 1 2 9
```

3.
```
   7 1, 8 7 8
  − 6 2, 3 7 6
```

4.
```
   4 7 9, 4 9 7
  −   7 8, 8 1 2
```

Lesson 2: Create conversion tables for length, weight, and capacity units using measurement tools, and use the tables to solve problems.

Name _____   Date _____

Practice Set C Part 1: Multi-Digit Subtraction with Zeros Fluency

1.
```
    7, 8 9 0
  − 5, 4 7 2
```

2.
```
   2 8, 0 0 1
  −    5, 8 5 3
```

3.
```
   6 0, 4 0 7
  − 3 5, 3 4 4
```

4.
```
   4 0 0, 0 6 9
  −    2 4, 3 6 2
```

Practice Set C Part 2: Multi-Digit Subtraction with Zeros Fluency

1.
```
    7, 8 9 0
  − 5, 4 7 2
```

2.
```
   2 8, 6 0 9
  −    6, 4 6 1
```

3.
```
   6 0, 4 9 7
  − 3 5, 4 3 4
```

4.
```
   4 0 0, 8 6 9
  −    2 5, 1 6 2
```

Lesson 2: Create conversion tables for length, weight, and capacity units using measurement tools, and use the tables to solve problems.

A STORY OF UNITS **Lesson 2 Core Fluency Practice Set D** 4•7

Name _____ Date _____

Practice Set D Part 1: Multi-Digit Addition and Subtraction Fluency

1.
```
      9, 3 2 7
  +   9, 6 6 4
```

2.
```
      3 9, 4 6 3
  -   3 8, 9 3 8
```

3.
```
      7 5 8, 1 9 4
  +       3 5, 4 7 8
```

4.
```
      8 3 9, 0 1 4
  -       2 7, 0 7 5
```

5.
```
      4 3 8, 6 1 5
  +   1 9 3, 9 7 9
```

6.
```
      9 6 0, 0 4 3
  -   3 6 8, 9 7 2
```

Practice Set D Part 2: Multi-Digit Addition and Subtraction Fluency

1.
```
      9, 6 3 0
  +   9, 3 6 1
```

2.
```
      3 4, 4 7 8
  -   3 3, 9 5 3
```

3.
```
      7 5 4, 4 5 4
  +       3 9, 2 1 8
```

4.
```
      8 3 9, 0 9 9
  -       2 7, 1 6 0
```

5.
```
      1 0 8, 2 1 5
  +   5 2 4, 3 7 9
```

6.
```
      9 5 9, 9 4 3
  -   3 6 8, 8 7 2
```

Lesson 2: Create conversion tables for length, weight, and capacity units using measurement tools, and use the tables to solve problems.

Name _____   Date _____

a.
Gallons	Quarts
1	
2	
3	
4	
5	
6	
7	
8	
9	
10	

The rule for converting gallons to quarts is _____.

b.
Quarts	Pints
1	
2	
3	
4	
5	
6	
7	
8	
9	
10	

The rule for converting quarts to pints is _____.

c.
Pints	Cups
1	
2	
3	
4	
5	
6	
7	
8	
9	
10	

d. 1 gallon = ____ pints

1 quart = ____ cups

1 gallon = ____ cups

The rule for converting pints to cups is _____.

A STORY OF UNITS   Lesson 2 Problem Set 4•7

Name _____   Date _____

Use RDW to solve Problems 1–3.

1. Susie has 3 quarts of milk. How many pints does she have?

2. Kristin has 3 gallons 2 quarts of water. Alana needs the same amount of water but only has 8 quarts. How many more quarts of water does Alana need?

3. Leonard bought 4 liters of orange juice. How many milliliters of juice does he have?

4. Complete the following conversion tables and write the rule under each table.

a.

Gallons	Quarts
1	
3	
5	
10	
13	

The rule for converting gallons to quarts is _____.

b.

Quarts	Pints
1	
2	
6	
10	
16	

The rule for converting quarts to pints is _____.

Lesson 2: Create conversion tables for length, weight, and capacity units using measurement tools, and use the tables to solve problems.

5. Solve.

   a. 8 gallons 2 quarts = _____ quarts

   b. 15 gallons 2 quarts = _____ quarts

   c. 8 quarts 2 pints = _____ pints

   d. 12 quarts 3 pints = _____ cups

   e. 26 gallons 3 quarts = _____ pints

   f. 32 gallons 2 quarts = _____ cups

6. Answer true or false for the following statements. If your answer is false, make the statement true.

   a. 1 gallon > 4 quarts   _____

   b. 5 liters = 5,000 milliliters   _____

   c. 15 pints < 1 gallon 1 cup   _____

7. Russell has 5 liters of a certain medicine. If it takes 2 milliliters to make 1 dose, how many doses can he make?

8. Each month, the Moore family drinks 16 gallons of milk and the Siler family goes through 44 quarts of milk. Which family drinks more milk each month?

9. Keith's lemonade stand served lemonade in glasses with a capacity of 1 cup. If he had 9 gallons of lemonade, how many cups could he sell?

A STORY OF UNITS   Lesson 2 Exit Ticket 4•7

Name _____ Date _____

1. Complete the table.

Quarts	Cups
1	
2	
4	

2. Bonnie's doctor recommended that she drink 2 cups of milk per day. If she buys 3 quarts of milk, will it be enough milk to last 1 week? Explain how you know.

A STORY OF UNITS **Lesson 2 Homework** 4•7

Name _____ Date _____

Use the RDW process to solve Problems 1–3.

1. Dawn needs to pour 3 gallons of water into her fish tank. She only has a 1-cup measuring cup. How many cups of water should she put in the tank?

2. Julia has 4 gallons 2 quarts of water. Ally needs the same amount of water but only has 12 quarts. How much more water does Ally need?

3. Sean drank 2 liters of water today, which was 280 milliliters more than he drank yesterday. How much water did he drink yesterday?

4. Complete the tables.

   a.

Gallons	Quarts
1	
2	
4	
12	
15	

   b.

Quarts	Pints
1	
2	
6	
10	
16	

Lesson 2: Create conversion tables for length, weight, and capacity units using measurement tools, and use the tables to solve problems.

A STORY OF UNITS   Lesson 2 Homework   4•7

5. Solve.

   a. 6 gallons 3 quarts = _____ quarts

   b. 12 gallons 2 quarts = _____ quarts

   c. 5 quarts 1 pint = _____ pints

   d. 13 quarts 3 pints = _____ cups

   e. 17 gallons 2 quarts = _____ pints

   f. 27 gallons 3 quarts = _____ cups

6. Explain how you solved Problem 5(f).

7. Answer true or false for the following statements. If your answer is false, make the statement true by correcting the right side of the comparison.

   a. 2 quarts > 10 pints   _____

   b. 6 liters = 6,000 milliliters _____

   c. 16 cups < 4 quarts 1 cup   _____

8. Joey needs to buy 3 quarts of chocolate milk. The store only sells it in pint containers. How many pints of chocolate milk should he buy? Explain how you know.

9. Granny Smith made punch. She used 2 pints of ginger ale, 3 pints of fruit punch, and 1 pint of orange juice. She served the punch in glasses that had a capacity of 1 cup. How many cups can she fill?

Lesson 3

Objective: Create conversion tables for units of time, and use the tables to solve problems.

Suggested Lesson Structure

- ■ Fluency Practice   (8 minutes)
- ■ Concept Development   (42 minutes)
- ■ Student Debrief   (10 minutes)
- **Total Time**   **(60 minutes)**

Fluency Practice (8 minutes)

- Grade 4 Core Fluency Differentiated Practice Sets **4.NBT.4**   (4 minutes)
- Convert Capacity Units **4.MD.1**   (4 minutes)

Grade 4 Core Fluency Differentiated Practice Sets (4 minutes)

Materials: (S) Core Fluency Practice Sets (Lesson 2 Core Fluency Practice Sets)

Note: During Topic A and for the remainder of the year, each day's Fluency Practice may include an opportunity for mastery of the addition and subtraction algorithm by means of the Core Fluency Practice Sets. The process is detailed and materials are provided in Lesson 2.

Convert Capacity Units (4 minutes)

Materials: (S) Personal white board

Note: This fluency activity reviews Lesson 2 and metric conversions from Module 2.

    T: (Write 1 L = ___ mL.) How many milliliters are in 1 liter?
    S: 1,000 milliliters.

Repeat the process for 2 and 3 liters.

    T: (Write 5 L = ___ mL.) Write the number sentence.
    S: (Write 5 L = 5,000 mL.)
    T: (Write 1 gal = ___ qt.) How many quarts are in 1 gallon?
    S: 4 quarts.

Repeat the process for 2 and 3 gallons.

    T: (Write 9 gal = ___ qt.) Write the number sentence.
    S: (Write 9 gal = 36 qt.)

A STORY OF UNITS   Lesson 3   4•7

T: (Write 1 qt = __ pt.) How many pints are in 1 quart?
S: 2 pints.

Repeat the process for 2 and 3 quarts.

T: (Write 7 qt = __ pt.) Write the number sentence.
S: (Write 7 qt = 14 pt.)
T: (Write 1 pt = __ c.) How many cups are in 1 pint?
S: 2 cups.

Repeat the process for 2 and 3 pints.

T: (Write 6 pt = __ c.) Write the number sentence.
S: (Write 6 pt = 12 c.)

Concept Development (42 minutes)

Materials: (T) Analog clock with a second hand, stopwatch
(S) Stopwatch (1 per group), personal white board, Practice Sheet

Problem 1: Convert minutes to seconds.

T: (Hold up an analog clock.) This clock has three hands. What units do the three hands count?
S: Hours, minutes, and seconds.
T: How many seconds are in 1 minute?
S: 60 seconds.
T: In your groups, one person will need to be in charge of the stopwatch. The rest of the group will close their eyes and, when they think 1 minute has passed, will write M on their personal white boards. The person with the stopwatch tells the group when 1 minute has passed. Let's see who comes closest to 1 minute.
T: (Allow students to finish the activity.) Did any of you write M on your board at exactly 1 minute?
T: (Display the two-column table.) What is the rule for converting minutes to seconds?
S: Multiply by 60.
T: Complete the conversion table for minutes to seconds.
S: (Complete the table.)
T: Solve for how many seconds are in 16 minutes.
S: 10 minutes is 600 seconds. 6 minutes is 360 seconds. 600 + 360 = 960. 960 seconds are in 16 minutes. → 16 times 60 is 960. 16 minutes equals 960 seconds.

Repeat the process with 23 minutes.

NOTES ON MATERIALS:

If stopwatches are not abundant enough to easily provide one for each group, consider virtual materials, such as those found at the following links:
Online Stopwatch: Fireworks
Online Stopwatch: Egg Timer

Minutes	Seconds
1	60
2	120
3	180
4	240
5	300
6	360
7	420
8	480
9	540
10	600

A STORY OF UNITS   Lesson 3   4•7

Problem 2: Convert hours to minutes.

T: Let's imagine, as we practiced in Grade 3, that we unwrap the clock and look at the numbers on a number line.

T: We can use the number line to help us determine the number of minutes in one hour. How many minutes are in an hour?

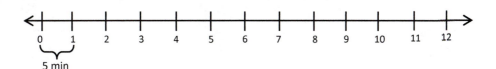

S: 60 minutes.

T: How many groups of 5 minutes?

S: 12 groups.

T: (Display the two-column table.) Complete the conversion table for hours to minutes.

T: How many minutes are in 18 hours?

S: 600 + 480 = 1,080. There are 1,080 minutes in 18 hours. → 18 × 60 = 1,080. 18 hours equals 1,080 minutes.

MP.8

T: How many seconds are in 18 minutes? (Pause.)

S: It's the same number! 1,080 seconds.

T: Why?

S: We multiplied by a factor of 60 for both. → The rule is times 60.

Hours	Minutes
1	60
2	120
3	180
4	240
5	300
6	360
7	420
8	480
9	540
10	600

Repeat the process with 36 hours.

Problem 3: Convert days to hours.

T: With your partner, determine the number of hours in 1 day. Use the number line if it helps you.

S: There are 24 hours in 1 day. The number line represents 12 hours, but I know that we need to double that because 12:00 a.m. to 12:00 p.m. is just half the day. A full day would be another 12 hours back to 12:00 a.m.

T: If we know that there are 24 hours in 1 day, we can complete the conversion table for days to hours. Complete the table.

T: How many hours are in 14 days?

S: 240 hours + 96 hours = 336 hours. → 14 × 24 = 336. 14 days equals 336 hours.

Days	Hours
1	24
2	48
3	72
4	96
5	120
6	144
7	168
8	192
9	216
10	240

Repeat the process for 42 days.

Lesson 3: Create conversion tables for units of time, and use the tables to solve problems.

Problem 4: Solve a word problem involving converting days to hours.

T: The Apollo 17 mission was completed in 12 days, 14 hours. How many hours did the mission last?

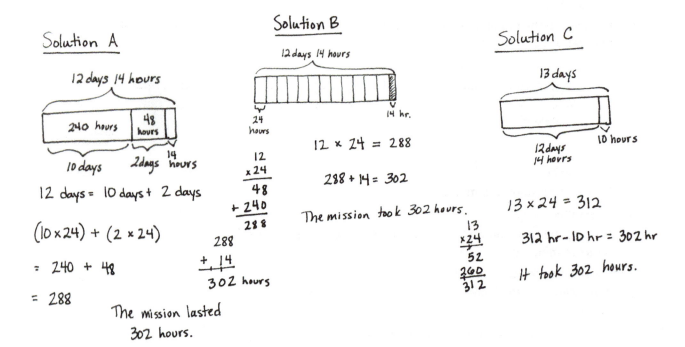

Problem Set (10 minutes)

Students should do their personal best to complete the Problem Set within the allotted 10 minutes. For some classes, it may be appropriate to modify the assignment by specifying which problems they work on first. Some problems do not specify a method for solving. Students should solve these problems using the RDW approach used for Application Problems.

NOTES ON MULTIPLE MEANS OF ENGAGEMENT:

Differentiate the difficulty of the word problem by offering students working above grade level these extension questions:

- Research the duration of Apollo 16's mission in days and hours. How many hours did it last?
- Compare the duration of Apollo 16 and 17's missions.

A STORY OF UNITS   Lesson 3   4•7

Student Debrief (10 minutes)

Lesson Objective: Create conversion tables for units of time, and use the tables to solve problems.

The Student Debrief is intended to invite reflection and active processing of the total lesson experience.

Invite students to review their solutions for the Problem Set. They should check work by comparing answers with a partner before going over answers as a class. Look for misconceptions or misunderstandings that can be addressed in the Debrief. Guide students in a conversation to debrief the Problem Set and process the lesson.

Any combination of the questions below may be used to lead the discussion.

- Explain how you could solve Problem 1 without a number line.
- Would it make sense to solve Problem 2 in seconds? Why or why not?
- Explain two strategies for solving problems converting a number of days to hours. Which method is most efficient and why? Which strategy did you use to solve Problem 7?
- Can anyone describe how time is kept in the military or in a foreign country? Is time (seconds, minutes, hours) defined as a metric or customary system?

Exit Ticket (3 minutes)

After the Student Debrief, instruct students to complete the Exit Ticket. A review of their work will help with assessing students' understanding of the concepts that were presented in today's lesson and planning more effectively for future lessons. The questions may be read aloud to the students.

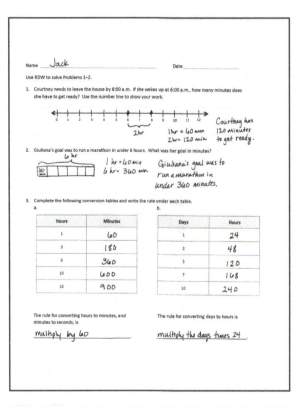

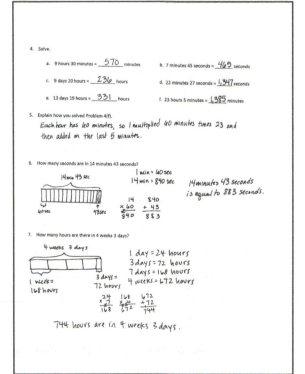

Lesson 3: Create conversion tables for units of time, and use the tables to solve problems.

47

Name _____ Date _____

a.

Minutes	Seconds
1	
2	
3	
4	
5	
6	
7	
8	
9	
10	

The rule for converting minutes to seconds is _____.

b.

Hours	Minutes
1	
2	
3	
4	
5	
6	
7	
8	
9	
10	

The rule for converting hours to minutes is _____.

c.

Days	Hours
1	
2	
3	
4	
5	
6	
7	
8	
9	
10	

The rule for converting days to hours is _____.

Name _____ Date _____

Use RDW to solve Problems 1–2.

1. Courtney needs to leave the house by 8:00 a.m. If she wakes up at 6:00 a.m., how many minutes does she have to get ready? Use the number line to show your work.

2. Giuliana's goal was to run a marathon in under 6 hours. What was her goal in minutes?

3. Complete the following conversion tables and write the rule under each table.

a.

Hours	Minutes
1	
3	
6	
10	
15	

The rule for converting hours to minutes and minutes to seconds is

_____.

b.

Days	Hours
1	
2	
5	
7	
10	

The rule for converting days to hours is

_____.

Lesson 3: Create conversion tables for units of time, and use the tables to solve problems.

4. Solve.

   a. 9 hours 30 minutes = _____ minutes

   b. 7 minutes 45 seconds = _____ seconds

   c. 9 days 20 hours = _____ hours

   d. 22 minutes 27 seconds = _____ seconds

   e. 13 days 19 hours = _____ hours

   f. 23 hours 5 minutes = _____ minutes

5. Explain how you solved Problem 4(f).

6. How many seconds are in 14 minutes 43 seconds?

7. How many hours are there in 4 weeks 3 days?

A STORY OF UNITS   Lesson 3 Exit Ticket   4•7

Name _____   Date _____

The astronauts from Apollo 17 completed 3 spacewalks while on the moon for a total duration of 22 hours 4 minutes. How many minutes did the astronauts walk in space?

Lesson 3: Create conversion tables for units of time, and use the tables to solve problems.

A STORY OF UNITS Lesson 3 Homework 4•7

Name _____ Date _____

Use RDW to solve Problems 1–2.

1. Jeffrey practiced his drums from 4:00 p.m. until 7:00 p.m. How many minutes did he practice? Use the number line to show your work.

2. Isla used her computer for 5 hours over the weekend. How many minutes did she spend on the computer?

3. Complete the following conversion tables and write the rule under each table.

a.

Hours	Minutes
1	
2	
5	
9	
12	

The rule for converting hours to minutes is _____.

b.

Days	Hours
1	
3	
6	
8	
20	

The rule for converting days to hours is _____.

Lesson 3: Create conversion tables for units of time, and use the tables to solve problems.

4. Solve.

   a. 10 hours 30 minutes = _____ minutes

   b. 6 minutes 15 seconds = _____ seconds

   c. 4 days 20 hours = _____ hours

   d. 3 minutes 45 seconds = _____ seconds

   e. 23 days 21 hours = _____ hours

   f. 17 hours 5 minutes = _____ minutes

5. Explain how you solved Problem 4(f).

6. It took a space shuttle 8 minutes 36 seconds to launch and reach outer space. How many seconds did it take?

7. Apollo 16's mission lasted just over 1 week 4 days. How many hours are there in 1 week 4 days?

Lesson 4

Objective: Solve multiplicative comparison word problems using measurement conversion tables.

Suggested Lesson Structure

■ Fluency Practice   (11 minutes)
■ Concept Development   (40 minutes)
■ Student Debrief   (9 minutes)
Total Time   **(60 minutes)**

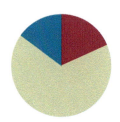

Fluency Practice (11 minutes)

- Grade 4 Core Fluency Differentiated Practice Sets   **4.NBT.4**   (4 minutes)
- Convert Length Units   **4.MD.1**   (4 minutes)
- Convert Weight Units   **4.MD.1**   (3 minutes)

Grade 4 Core Fluency Differentiated Practice Sets (4 minutes)

Materials: (S) Core Fluency Practice Sets (Lesson 2 Core Fluency Practice Sets)

Note: During Topic A and for the remainder of the year, each day's Fluency Practice may include an opportunity for mastery of the addition and subtraction algorithm by means of the Core Fluency Practice Sets. The process is detailed and Practice Sets are provided in Lesson 2.

Convert Length Units (4 minutes)

Materials: (S) Personal white board

Note: This fluency activity reviews Lesson 1 and metric conversions from Module 2.

    T: (Write 1,000 m.) 1,000 m is the same as 1 of what unit?
    S: 1 kilometer.
    T: (Write 1,000 m = 1 km.)

Repeat the process for 2,000 and 3,000 meters.

    T: (Write 6,000 m = ___ km.) Write the number sentence.
    S: (Write 6,000 m = 6 km.)

T: (Write 100 cm.) 100 cm is the same as 1 of what unit?
S: 1 meter.
T: (Write 100 cm = 1 m.)

Repeat the process for 200 and 300 meters.

T: (Write 700 cm = ___ m.) Write the number sentence.
S: (Write 700 cm = 7 m.)
T: (Write 3 ft.) 3 feet is the same as 1 of what unit?
S: 1 yard.
T: (Write 3 ft = 1 yd.)

Repeat the process for 6 and 9 yards.

T: (Write 21 ft = ___ yd.) Write the number sentence.
S: (Write 21 ft = 7 yd.)
T: (Write 12 in.) 12 inches is the same as 1 of what unit?
S: 1 foot.
T: (Write 12 in = 1 ft.)

Repeat the process for 24 and 36 inches.

Convert Weight Units (3 minutes)

Materials: (T) Personal white board

Note: This fluency activity reviews Lesson 1 and metric conversions from Module 2.

T: (Write 1,000 g.) 1,000 g is the same as 1 of what unit?
S: 1 kg.
T: (Write 1,000 g = 1 kg.)

Repeat the process for 2,000 and 3,000 grams.

T: (Write 16 oz.) 16 ounces is the same as 1 of what unit?
S: 1 pound.
T: (Write 16 oz = 1 lb.)

Repeat the process for 32 and 48 ounces.

Concept Development (40 minutes)

Materials: (S) Problem Set

Suggested Delivery of Instruction for Solving Lesson 4's Word Problems

For Problems 1–5, students may work in pairs to solve each of the problems using the RDW approach to problem solving. These problems are also found in the Problem Set.

1. Model the problem.

Select two pairs of students who can be successful with modeling the problem to work at the board while the other students work independently or in pairs at their seats. Review the following questions before beginning the first problem.

- Can you draw something?
- What can you draw?
- What conclusions can you make from your drawing?

As students work, circulate. Reiterate the questions above.

After two minutes, have the two pairs of students share only their labeled diagrams.

For about one minute, have the demonstrating students receive and respond to feedback and questions from their peers.

2. Calculate to solve, and write a statement.

Allow students two minutes to complete work on the problem, sharing their work and thinking with a peer. Have students write their equations and statements of the answer.

3. Assess the solution.

Give students one to two minutes to assess the solutions presented by their peers on the board, comparing the solutions to their own work. Highlight alternative methods to reach the correct solution.

Problem 1

Beth is allowed 2 hours of TV time each week. Her sister is allowed 2 times as much. How many minutes of TV can Beth's sister watch?

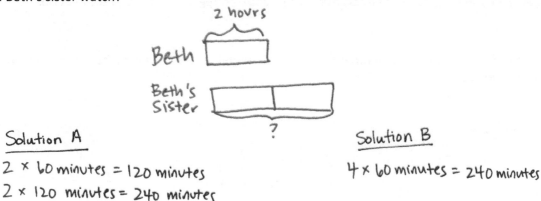

Solution A

2 × 60 minutes = 120 minutes
2 × 120 minutes = 240 minutes

Solution B

4 × 60 minutes = 240 minutes

Beth's sister is allowed to watch 240 minutes of TV each week.

This two-step problem requires students to determine the number of hours Beth's sister is allowed to watch TV and then use that information to determine the time in minutes. In Solution A, students solve for the number of minutes in 1 unit by multiplying 60 minutes by 2. Then, they multiply 120 minutes by 2 to solve for the number of minutes Beth's sister watches TV. In Solution B, students recognize 2 units as 4 hours, multiplying 4 by 60 minutes to solve for 240 minutes.

Problem 2

Clay weighs 9 times as much as his baby sister. Clay weighs 63 pounds. How much does his baby sister weigh in ounces?

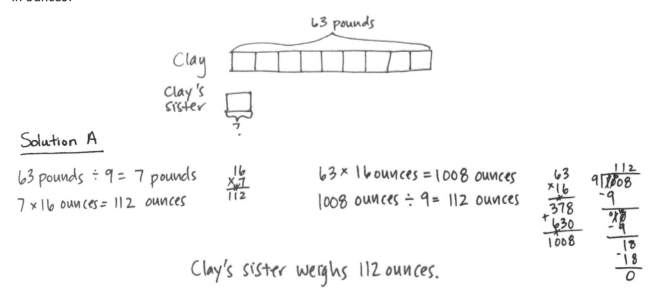

In Solution A, students first determine Clay's sister's weight in pounds by dividing by 9. Then, they convert pounds to ounces to get 112 ounces. In Solution B, students determine Clay's weight in ounces first. Next, they divide his weight in ounces by 9 to solve for 112 ounces. Notice that Solution A is more efficient and requires fewer paper-and-pencil calculations overall, but both strategies reach a correct solution.

Problem 3

Helen has 4 yards of rope. Daniel has 4 times as much rope as Helen. How many more feet of rope does Daniel have than Helen?

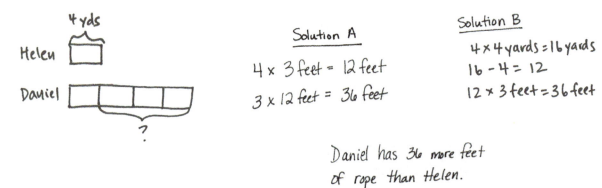

In Solution A, students convert Helen's rope from yards to feet first. Then they multiply by 3 to find the value of 3 units; the difference is clearly shown in the model. In Solution B, students determine the number of yards in Daniel's rope by first multiplying by 4. They subtract the length of Helen's rope from Daniel's rope to find the difference. Finally, they convert 12 yards to feet by multiplying by 3. Again, have students notice the greater efficiency of Solution A.

NOTES ON MULTIPLE MEANS OF REPRESENTATION:

In addition to the tape diagram, learners can construct a conversion table to solve Problem 3, such as the following:

Liters	Milliliters
1	1,000
5	5,000
10	10,000
50	

Lesson 4: Solve multiplicative comparison word problems using measurement conversion tables.

Problem 4

A dishwasher uses 11 liters of water for each cycle. A washing machine uses 5 times as much water as a dishwasher uses for each load. Combined, how many milliliters of water are used for 1 cycle of each machine?

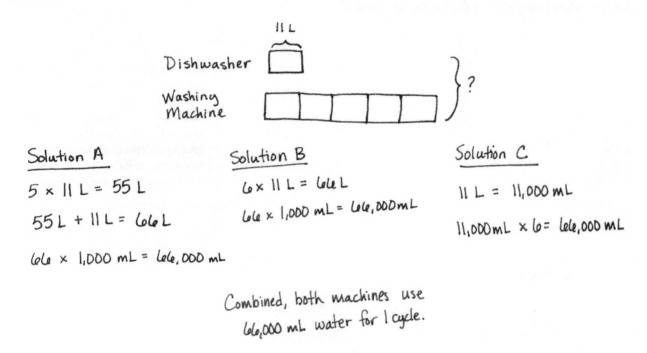

Solution A allows students to determine the amount of water used by the washing machine first by multiplying 11 liters by 5 to get 55 liters and then by adding the water used by the dishwasher, 11 liters, to find that both machines use 66 liters of water. Finally, students convert liters to milliliters by multiplying by 1,000. In Solution B, students solve for a total of 6 units, multiplying 11 liters times 6. Next, students convert to milliliters by multiplying by 1,000. In Solution C, the number of milliliters in 11 liters is found first and then multiplied by 6.

A STORY OF UNITS   Lesson 4   4•7

Problem 5

Joyce bought 2 pounds of apples. She bought 3 times as many pounds of potatoes as pounds of apples. The melons she bought were 10 ounces lighter than the total weight of the potatoes. How many ounces did the melons weigh?

NOTES ON MULTIPLE MEANS OF ACTION AND EXPRESSION:

Students working above grade level may enjoy an open-ended independent challenge. As an alternative to solving Problem 5, students may use the unlabeled tape diagram (as pictured) to write their own word problems. In keeping with the objective, students should include measurement conversions in their word problems.

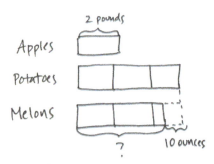

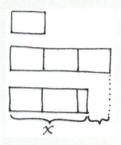

MP.2

Solution A

Potatoes:
3 × 2 pounds = 6 pounds
6 × 16 ounces = 96 ounces

96 − 10 = 86

Solution B

Apples:
2 × 16 ounces = 32 ounces
Potatoes:
3 × 32 ounces = 96 ounces
Melons:
96 − 10 = 86

The melons weigh 86 ounces.

In Solution A, students first find the weight of the potatoes in pounds by multiplying by 3 and then convert the weight of the potatoes to ounces by multiplying by 16. They then subtract 10 ounces from the potatoes to get 86 ounces. With Solution B, students convert the weight of the apples to ounces first by multiplying by 16. Then, they determine the weight of the potatoes in ounces by multiplying by 3. Finally, they subtract 10 from 96 to get 86 ounces. Look for other plausible solutions, such as solving for the 2 units of melons and adding on 22 ounces of the next partial unit. Have students assess the reasonableness of their answers by seeing that if their tape diagrams model the weight of the melons as being less than the potatoes but greater than the apples, then their answers also show, paying close attention to the units, that they are comparing.

Lesson 4:   Solve multiplicative comparison word problems using measurement conversion tables.

A STORY OF UNITS   Lesson 4   4•7

Problem Set

Please note that the Problem Set is completed as part of the Concept Development for this lesson.

Student Debrief (9 minutes)

Lesson Objective: Solve multiplicative comparison word problems using measurement conversion tables.

The Student Debrief is intended to invite reflection and active processing of the total lesson experience.

Invite students to review their solutions for the Problem Set. They should check work by comparing answers with a partner before going over answers as a class. Look for misconceptions or misunderstandings that can be addressed in the Debrief. Guide students in a conversation to debrief the Problem Set and process the lesson.

Any combination of the questions below may be used to lead the discussion.

- Share your strategy for solving Problem 3 with your partner. What did your partner do well? What could he or she have done differently?
- How were the set-ups for Problem 3 and Problem 4 similar to each other? How were they different?
- In today's problems, why do we always have to convert the units?
- At what point in solving Problem 5 did you choose to convert into ounces? Is it better to convert to ounces earlier or at the end? Why?

Exit Ticket (3 minutes)

After the Student Debrief, instruct students to complete the Exit Ticket. A review of their work will help with assessing students' understanding of the concepts that were presented in today's lesson and planning more effectively for future lessons. The questions may be read aloud to the students.

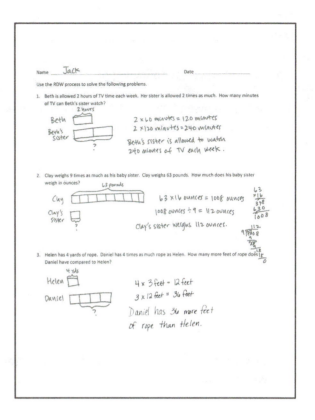

Lesson 4: Solve multiplicative comparison word problems using measurement conversion tables.

Name _____ Date _____

Use RDW to solve the following problems.

1. Beth is allowed 2 hours of TV time each week. Her sister is allowed 2 times as much. How many minutes of TV can Beth's sister watch?

2. Clay weighs 9 times as much as his baby sister. Clay weighs 63 pounds. How much does his baby sister weigh in ounces?

3. Helen has 4 yards of rope. Daniel has 4 times as much rope as Helen. How many more feet of rope does Daniel have compared to Helen?

4. A dishwasher uses 11 liters of water for each cycle. A washing machine uses 5 times as much water as a dishwasher uses for each load. Combined, how many milliliters of water are used for 1 cycle of each machine?

5. Joyce bought 2 pounds of apples. She bought 3 times as many pounds of potatoes as pounds of apples. The melons she bought were 10 ounces lighter than the total weight of the potatoes. How many ounces did the melons weigh?

A STORY OF UNITS

Lesson 4 Exit Ticket 4•7

Name _____ Date _____

Use RDW to solve the following problem.

Brian has a melon that weighs 3 pounds. He cut it into six equal pieces. How many ounces did each piece weigh?

A STORY OF UNITS

Lesson 4 Homework 4•7

Name _____ Date _____

Use RDW to solve the following problems.

1. Sandy took the train to New York City. The trip took 3 hours. Jackie took the bus, which took twice as long. How many minutes did Jackie's trip take?

2. Coleton's puppy weighed 3 pounds 8 ounces at birth. The vet weighed the puppy again at 6 months, and the puppy weighed 7 pounds. How many ounces did the puppy gain?

3. Jessie bought a 2-liter bottle of juice. Her sister drank 650 milliliters. How many milliliters were left in the bottle?

Lesson 4: Solve multiplicative comparison word problems using measurement conversion tables.

65

4. Hudson has a chain that is 1 yard in length. Myah's chain is 3 times as long. How many feet of chain do they have in all?

5. A box weighs 8 ounces. A shipment of boxes weighs 7 pounds. How many boxes are in the shipment?

6. Tracy's rain barrel has a capacity of 27 quarts of water. Beth's rain barrel has a capacity of twice the amount of water as Tracy's rain barrel. Trevor's rain barrel can hold 9 quarts of water less than Beth's barrel.

   a. What is the capacity of Trevor's rain barrel?

   b. If Tracy, Beth, and Trevor's rain barrels were filled to capacity, and they poured all of the water into a 30-gallon bucket, would there be enough room? Explain.

Lesson 5

Objective: Share and critique peer strategies.

Suggested Lesson Structure

- **Fluency Practice** (12 minutes)
- **Concept Development** (38 minutes)
- **Student Debrief** (10 minutes)

Total Time **(60 minutes)**

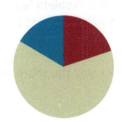

Fluency Practice (12 minutes)

- Sprint: Convert Length Units   4.MD.1   (9 minutes)
- Convert Time Units   4.MD.2   (3 minutes)

Sprint: Convert Length Units (9 minutes)

Materials: (S) Convert Length Units Sprint

Note: This fluency activity reviews Lesson 1.

Convert Time Units (3 minutes)

Note: This fluency activity reviews Lesson 3.

    T: (Write 1 hr = ___ min.) How many minutes are in 1 hour?
    S: 60 minutes.

Repeat the process with 2, 3, and 10 hours.

    T: (Write 1 min = ___ sec.) How many seconds are in 1 minute?
    S: 60 seconds.

Repeat the process with 2, 3, and 10 minutes.

A STORY OF UNITS   Lesson 5 4•7

Concept Development (38 minutes)

Materials: (S) Problem Set, peer share and critique form (Template)

Note: In Problem 1, students work in pairs to create a word problem to match the tape diagram and analyze the work using the share and critique form. The Problem Set is used in Problem 2 of the Concept Development for students to write and solve problems and then to share and critique strategies with their peers using the share and critique form.

Problem 1: Create and analyze student work using the share and critique form.

Display the following:

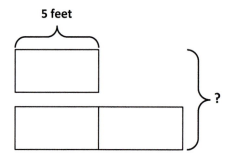

T: What information do we know by looking at this diagram?

S: One unit is 5 feet long. → The other tape is twice as long. → We are looking for the total value of all 3 units. → The value of the second tape diagram is 2 times the length of the first. It's 10 feet, so the value of both is 15 feet.

T: With your partner, think about a word problem that would go with this diagram. Take into consideration the units we are using when creating your word problem. Express your final answer in inches. (Answers will vary.)

Student A

Joe is 5 feet tall.
Bill is twice as tall.
How tall are they together?
2 × 5 feet = 10 feet
10 + 5 = 15
15 × 12 inches = 180 inches
Together they are 180 inches.

```
  15
× 12
  30
 150
 180
```

Student B

Mariana's kite has a string 5 feet long.
Heather's kite's string is twice as long.
How long are their strings all together?
5 feet × 3 = 15 feet
15 × 12 inches = 180 inches
Mariana and Heather's strings are 180 inches all together.

```
  15
× 12
  30
 150
 180
```

Lesson 5: Share and critique peer strategies.

68

Debrief the problem using questions from the share and critique form. Ensure students notice that in Solution A, it is not realistic for Bill to be 10 feet tall. Have them generate a more realistic modification.

Display:

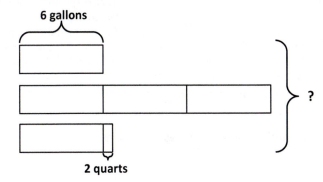

T: Turn to your partner, and share what information you know from looking at this diagram. Then, create a word problem that goes with this diagram to solve for the number of quarts.

> **NOTES ON MULTIPLE MEANS OF ACTION AND EXPRESSION:**
>
> English language learners may need scaffolds for writing word problems. Provide a word bank or sentence frames, and allow students to discuss their thoughts before writing. Some possible sentence frames are given below.
>
> - ___ had 6 gallons of ___ .
> - ___ had 3 times as much.
> - ___ had 2 quarts more than ___ .

Student A

For the holiday party, the Kindergarten had 6 gallons of juice. The first grade had 3 times as much juice as the Kindergarten. The second grade had 2 quarts more juice than the Kindergarten. How much juice did the Kindergarten, first grade, and second grade have all together?

K: 6 gal
1st: 3 × 6 gal = 18 gal
2nd: 6 gal 2 qt
6 gal + 18 gal + 6 gal = 30 gal
30 × 4 qt = 120 qt
120 qt + 2 qt = 122 qt

The Kindergarten, 1st, and 2nd grade classes had 122 quarts of juice all together.

Student B

Peter filled his car with 6 gallons of gasoline. Doug filled his car with 3 times as much gas as Peter filled his car with. Wesley filled his car with 2 quarts gas more than the amount of gas Peter filled his car with. What is the total number of quarts of gas filled into the three cars?

6 × 4 qt = 24 qt
3 × 24 qt = 72 qt
24 qt + 2 qt = 26 qt
24 qt + 72 qt + 26 qt = 122 qt

122 quarts of gas were filled into the three cars.

Lesson 5: Share and critique peer strategies.

Debrief the problem using questions from the share and critique form. Students might notice that gasoline is often measured in gallons rather than quarts. However, it is not unrealistic or wrong to state the capacity in quarts.

NOTES ON MULTIPLE MEANS OF ENGAGEMENT:

In the interest of cultivating a non-threatening atmosphere in which all students feel confident and comfortable sharing and receiving feedback, you may want to discuss goals, set guidelines, and model positive giving and receiving of balanced feedback.

Problem 2: Share and critique work.

Distribute the Problem Set and share and critique form.

T: Work with a partner to complete the Problem Set. When you are finished solving and creating a word problem to go along with each diagram, turn to your partner and share. Use the peer share and critique form to take notes about your work and your partner's work.

Student equations and responses will vary. Circulate and assist students as necessary.

1. a. Label the rest of the tape diagram below. Solve for the unkown.

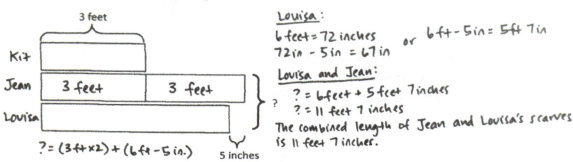

b. Write your own problem that could be solved using the diagram above.

Kit knitted a scarf that was 3 feet long.
Jean knitted a scarf two times as long as Kit's.
Louisa's scarf was 5 inches shorter than Jean's.
How long were Jean and Louisa's scarves combined?

---

2. Create your own problem using the diagram below and solve for the unknown.

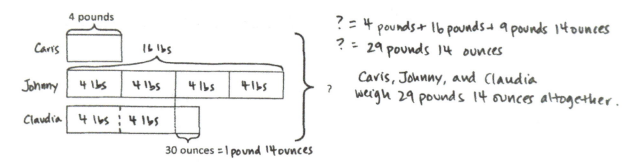

Caris weighs 4 pounds.
Johnny weighs four times as much as Caris.
Claudia weighs 30 ounces more than half of Johnny's weight.
How much do they weigh altogether?

A STORY OF UNITS   Lesson 5   4•7

Problem Set

Please note that the Problem Set is completed as part of the second half of the Concept Development for this lesson.

Student Debrief (10 minutes)

Lesson Objective: Share and critique peer strategies.

The Student Debrief is intended to invite reflection and active processing of the total lesson experience.

Invite students to review their solutions for the Problem Set. They should check work by comparing answers with a partner before going over answers as a class. Look for misconceptions or misunderstandings that can be addressed in the Debrief. Guide students in a conversation to debrief the Problem Set and process the lesson.

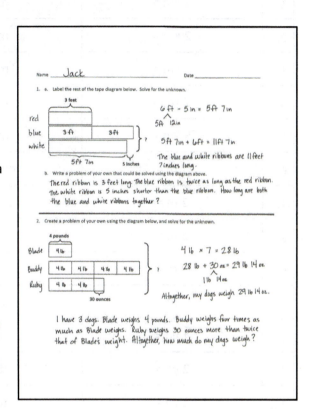

Any combination of the questions below may be used to lead the discussion.

- What did your partner do well in solving Problem 1? Problem 2?
- Did you and your partner use different strategies to solve Problem 1? If so, how were they different?
- How might you improve your work in solving and creating a word problem for Problem 2?
- What do we learn by analyzing different strategies for solving a problem?

Exit Ticket (3 minutes)

After the Student Debrief, instruct students to complete the Exit Ticket. A review of their work will help with assessing students' understanding of the concepts that were presented in today's lesson and planning more effectively for future lessons. The questions may be read aloud to the students.

Lesson 5:   Share and critique peer strategies.   71

A STORY OF UNITS   Lesson 5 Sprint 4•7

A
Number Correct: _____

Convert Length Units

1.	1 km =	m	23.	6 km =	m
2.	2 km =	m	24.	5 m =	cm
3.	3 km =	m	25.	7 m =	cm
4.	7 km =	m	26.	4 m =	cm
5.	5 km =	m	27.	8 m =	cm
6.	1 m =	cm	28.	4 yd =	ft
7.	2 m =	cm	29.	8 yd =	ft
8.	3 m =	cm	30.	6 yd =	ft
9.	9 m =	cm	31.	9 yd =	ft
10.	6 m =	cm	32.	5 ft =	in
11.	1 yd =	ft	33.	6 ft =	in
12.	2 yd =	ft	34.	1,000 m =	km
13.	3 yd =	ft	35.	8,000 m =	km
14.	10 yd =	ft	36.	100 cm =	m
15.	5 yd =	ft	37.	600 cm =	m
16.	1 ft =	in	38.	3 ft =	yd
17.	2 ft =	in	39.	24 ft =	yd
18.	3 ft =	in	40.	12 in =	ft
19.	10 ft =	in	41.	72 in =	ft
20.	4 ft =	in	42.	8 ft =	in
21.	9 km =	m	43.	84 in =	ft
22.	4 km =	m	44.	9 ft =	in

Lesson 5:   Share and critique peer strategies.

B

Convert Length Units

Number Correct: _____
Improvement: _____

#			#		
1.	1 m =	cm	23.	6 m =	cm
2.	2 m =	cm	24.	5 km =	m
3.	3 m =	cm	25.	7 km =	m
4.	7 m =	cm	26.	4 km =	m
5.	5 m =	cm	27.	8 km =	m
6.	1 km =	m	28.	6 yd =	ft
7.	2 km =	m	29.	9 yd =	ft
8.	3 km =	m	30.	4 yd =	ft
9.	9 km =	m	31.	8 yd =	ft
10.	6 km =	m	32.	5 ft =	in
11.	1 yd =	ft	33.	6 ft =	in
12.	2 yd =	ft	34.	100 cm =	m
13.	3 yd =	ft	35.	800 cm =	m
14.	5 yd =	ft	36.	1,000 m =	km
15.	10 yd =	ft	37.	6,000 m =	km
16.	1 ft =	in	38.	3 ft =	yd
17.	2 ft =	in	39.	27 ft =	yd
18.	3 ft =	in	40.	12 in =	ft
19.	10 ft =	in	41.	84 in =	ft
20.	4 ft =	in	42.	9 ft =	in
21.	9 m =	cm	43.	72 in =	ft
22.	4 m =	cm	44.	8 ft =	in

A STORY OF UNITS   Lesson 5 Problem Set 4•7

Name _____ Date _____

1. a. Label the rest of the tape diagram below. Solve for the unknown.

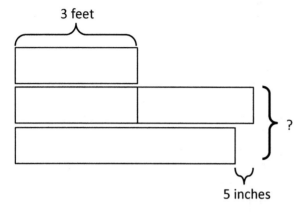

   b. Write a problem of your own that could be solved using the diagram above.

2. Create a problem of your own using the diagram below, and solve for the unknown.

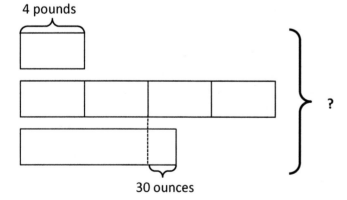

Lesson 5 Exit Ticket 4•7

Name _____   Date _____

Caitlin ran 1,680 feet on Monday and 2,340 feet on Tuesday. How many yards did she run in those two days?

Name _____ Date _____

Draw a tape diagram to solve the following problems.

1. Timmy drank 2 quarts of water yesterday. He drank twice as much water today as he drank yesterday. How many cups of water did Timmy drink in the two days?

2. Lisa recorded a 2-hour television show. When she watched it, she skipped the commercials. It took her 84 minutes to watch the show. How many minutes did she save by skipping the commercials?

3. Jason bought 2 pounds of cashews. Sarah ate 9 ounces. David ate 2 ounces more than Sarah. How many ounces were left in Jason's bag of cashews?

4. a. Label the rest of the tape diagram below. Solve for the unknown.

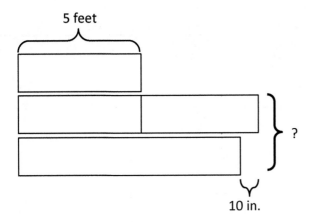

b. Write a problem of your own that could be solved using the diagram above.

5. Create a problem of your own using the diagram below, and solve for the unknown.

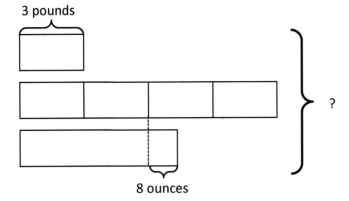

Classmate:		Problem Number:	
Strategies my classmate used:			
Things my classmate did well:			
Suggestions for improvement:			
Changes I would make to my work based on my classmate's work:			

Classmate:		Problem Number:	
Strategies my classmate used:			
Things my classmate did well:			
Suggestions for improvement:			
Changes I would make to my work based on my classmate's work:			

peer share and critique form

A STORY OF UNITS

Mathematics Curriculum

GRADE 4 • MODULE 7

Topic B
Problem Solving with Measurement

4.OA.2, 4.OA.3, 4.MD.1, 4.MD.2, 4.NBT.5, 4.NBT.6

Focus Standards:	4.OA.2	Multiply or divide to solve word problems involving multiplicative comparison, e.g., by using drawings and equations with a symbol for the unknown number to represent the problem, distinguishing multiplicative comparison from additive comparison. (See CCSS-M Glossary, Table 2.)
	4.OA.3	Solve multi-step word problems posed with whole numbers and having whole-number answers using the four operations, including problems in which remainders must be interpreted. Represent these problems using equations with a letter standing for the unknown quantity. Assess the reasonableness of answers using mental computation and estimation strategies including rounding.
	4.MD.1	Know relative sizes of measurement units within one system of units including km, m, cm; kg, g; lb, oz.; l, ml; hr, min, sec. Within a single system of measurement, express measurements in a larger unit in terms of a smaller unit. Record measurement equivalents in a two-column table. *For example, know that 1 ft is 12 times as long as 1 in. Express the length of a 4 ft snake as 48 in. Generate a conversion table for feet and inches listing the number pairs (1, 12), (2, 24), (3, 36), ...*
	4.MD.2	Use the four operations to solve word problems involving distances, intervals of time, liquid volumes, masses of objects, and money, including problems involving simple fractions or decimals, and problems that require expressing measurements given in a larger unit in terms of a smaller unit. Represent measurement quantities using diagrams such as number line diagrams that feature a measurement scale.
Instructional Days:	6	
Coherence -Links from:	G3–M1	Properties of Multiplication and Division and Solving Problems with Units of 2–5 and 10
	G3–M2	Place Value and Problem Solving with Units of Measure
-Links to:	G5–M1	Place Value and Decimal Fractions
	G5–M2	Multi-Digit Whole Number and Decimal Fraction Operations

Each lesson in Topic B builds upon the conversion work from Topic A to add and subtract mixed units of capacity, length, weight, and time. Unlike the mixed unit work in Module 2, now students work with the two systems of measurement, customary and metric, as well as being presented with fractional amounts of measurement, for example, $2\frac{3}{4}$ feet or $4\frac{3}{8}$ pounds. As students add like units, they make comparisons to adding like fractional units, further establishing the importance of deeply understanding the unit. Just as 2 fourths + 3 fourths = 5 fourths, so does 2 quarts + 3 quarts = 5 quarts. 5 fourths can be decomposed into

1 one 1 fourth; therefore; 5 quarts can be decomposed into 1 gallon 1 quart. Students realize that this also applies to subtraction: Just as $1 - \frac{3}{4}$ must be renamed to $\frac{4}{4} - \frac{3}{4}$ so the units are alike, the units of measurement must be renamed to make like units (1 quart – 3 cups = 4 cups – 3 cups). Students go on to add and subtract mixed units of measurements, finding multiple solution strategies, similar to the mixed number work in fractions.

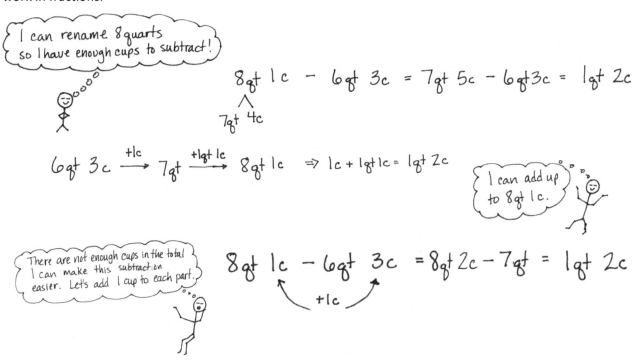

In Lessons 6–9, each lesson focuses on a specific type of measurement: capacity, length, weight, or time. Students go on to practice addition and subtraction of mixed units of measurements to solve multi-step word problems in Lessons 10 and 11.

Judy spent 1 hour and 15 minutes less than Sandy exercising last week. Sandy spent 50 minutes less than Mary, who spent 3 hours at the gym. How long did Judy exercise?

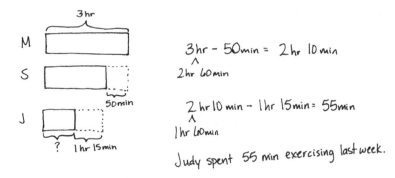

Topic B 4•7

A Teaching Sequence Toward Mastery of Problem Solving with Measurement
Objective 1: Solve problems involving mixed units of capacity. (Lesson 6)
Objective 2: Solve problems involving mixed units of length. (Lesson 7)
Objective 3: Solve problems involving mixed units of weight. (Lesson 8)
Objective 4: Solve problems involving mixed units of time. (Lesson 9)
Objective 5: Solve multi-step measurement word problems. (Lessons 10–11)

A STORY OF UNITS   Lesson 6   4•7

Lesson 6

Objective: Solve problems involving mixed units of capacity.

Suggested Lesson Structure

■ Fluency Practice	(12 minutes)
■ Concept Development	(36 minutes)
■ Student Debrief	(12 minutes)
Total Time	**(60 minutes)**

Fluency Practice (12 minutes)

- Grade 4 Core Fluency Differentiated Practice Sets **4.NBT.4**   (4 minutes)
- Add Mixed Numbers **4.NF.3c**   (4 minutes)
- Convert Capacity Units **4.MD.2**   (4 minutes)

Grade 4 Core Fluency Differentiated Practice Sets (4 minutes)

Materials: (S) Core Fluency Practice Sets (Lesson 2 Core Fluency Practice Sets)

Note: During Module 7, each day's Fluency Practice may include an opportunity for mastery of the addition and subtraction algorithm by means of the Core Fluency Practice Sets. The process is detailed and Practice Sets are provided in Lesson 2.

Add Mixed Numbers (4 minutes)

Materials: (S) Personal white board

Note: This fluency activity anticipates today's lesson by adding fractional units directly relevant to the measurement units within the lesson: $\frac{1}{2}$, $\frac{1}{4}$, and $\frac{1}{8}$. Direct students to respond chorally or with a written response.

T: 3 fourths + 2 fourths is how many fourths?
S: 5 fourths.
T: Express 5 fourths as ones and fourths.
S: 1 and 1 fourth.
T: 3 fourths + 3 fourths is how many fourths?
S: 6 fourths.

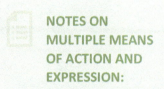

NOTES ON MULTIPLE MEANS OF ACTION AND EXPRESSION:

Fluency drills are fun, fast-paced math games, but English language learners may struggle to keep up. Some students may not understand how to respond to, "Express 5 fourths as ones and fourths." Provide an example, coupling language with visual aids or gestures, check for understanding, and, if necessary, explain in the students' first language.

T: Express 6 fourths as ones and fourths.
S: 1 and 2 fourths.

Continue with the following possible sequence: $\frac{1}{2} + \frac{4}{2}, \frac{3}{8} + \frac{7}{8}, \frac{5}{8} + \frac{6}{8}, \frac{7}{8} + \frac{7}{8}$.

Convert Capacity Units (4 minutes)

Note: This fluency activity reviews Lesson 2 and anticipates work with capacity units.

T: Express each number of gallons and quarts as quarts.
T: 1 gallon.
S: 4 quarts.
T: 1 gallon 1 quart.
S: 5 quarts.
T: 1 gallon 3 quarts.
S: 7 quarts.
T: 2 gallons.
S: 8 quarts.
T: Express each number of quarts as gallons and quarts if possible.
T: 4 quarts is …?
S: 1 gallon.
T: 8 quarts is …?
S: 2 gallons.

Repeat the process with quarts and pints and then gallons and pints.

Concept Development (36 minutes)

Materials: (S) Personal white board

Problem 1: Add mixed units of capacity.

T: 2 cats + 3 cats is …?
S: 5 cats.
T: 2 fourths + 3 fourths is …?
S: 5 fourths.
T: Express 5 fourths as a mixed number.
S: 1 and 1 fourth.
T: 2 quarts + 3 quarts is …?
S: 5 quarts.
T: Express 5 quarts as gallons and quarts.
S: 1 gallon 1 quart.

MP.7

Lesson 6: Solve problems involving mixed units of capacity.

83

A STORY OF UNITS — Lesson 6 4•7

T: Here are two different ways of adding 2 quarts and 3 quarts. Analyze them with your partner.

Solution A

$2 \text{ qt} \xrightarrow{+2\text{qt}} 1 \text{ gal} \xrightarrow{+1\text{qt}} 1 \text{ gal } 1 \text{ qt}$

Solution B

$2 \text{ qt} + 3 \text{ qt} = 5 \text{ qt} = 1 \text{ gal } 1 \text{ qt}$
(5 qt = 4 qt + 1 qt)

MP.7

S: Solution A makes 1 gallon first by adding on 2 quarts. → Solution B adds the quarts together and then takes out 1 gallon from 5 quarts. → Solution A completes a gallon just like if we were adding $\frac{2}{4}$ and $\frac{3}{4}$ and made one by adding $\frac{2}{4}$. → Solution B is like adding $\frac{2}{4}$ and $\frac{3}{4}$, getting $\frac{5}{4}$, and then taking out $\frac{4}{4}$ to get one and 1 fourth.

T: Yes, we can either complete a gallon first and then add on the remaining quarts or add to get 5 quarts and then rename to make 1 gallon and 1 quart.

> **NOTES ON MULTIPLE MEANS OF ENGAGEMENT:**
>
> Today's lesson of partner work and discussion fosters collaboration and communication that is valuable to students working below grade level because it may increase opportunities for one-on-one support and sustained engagement. Some learners may benefit from clear guidance in working effectively with others. Successful engagement comes by providing clear roles and responsibilities for partners or rubrics and norms that communicate partner work expectations.

Allow students to choose a method to solve and express the following sums with mixed units:

- 3 quarts + 3 quarts
- 2 cups + 3 cups
- 3 pints + 4 pints

T: Here are two different ways of adding 5 gallons 2 quarts + 3 quarts. Analyze them with a partner.

Solution C

$5 \text{ gal } 2 \text{ qt} \xrightarrow{+2\text{qt}} 6 \text{ gal} \xrightarrow{+1\text{qt}} 6 \text{ gal } 1 \text{ qt}$

Solution D

$5 \text{ gal } 2 \text{ qt} + 3 \text{ qt} = 5 \text{ gal } 5 \text{ qt} = 6 \text{ gal } 1 \text{ qt}$
(5 qt = 1 gal + 1 qt)

S: Solution C makes 1 gallon first by counting up 2 quarts to get 6 gallons and then adding on the extra quart. → Solution D adds the quarts together to get 5 gallons 5 quarts and then takes out one gallon from 5 quarts. → It's like adding mixed numbers—we add the like units.

Allow students to choose a method to solve and express the following sums with mixed units:

- 3 gallons 1 quart + 3 quarts
- 17 quarts 3 cups + 3 cups
- 4 gallons 7 pints + 7 pints

Lesson 6: Solve problems involving mixed units of capacity.

A STORY OF UNITS   Lesson 6 4•7

T: Here are two different ways of adding 5 gallons 2 quarts + 4 gallons 3 quarts. Analyze them with a partner.

Solution E

5 gal 2 qt →+4gal→ 9 gal 2 qt →+2qt→ 10 gal →+1qt→ 10 gal 1 qt

Solution F

5 gal 2 qt + 4 gal 3 qt = 9 gal 5 qt = 10 gal 1 qt
                                       /\
                                    1 gal 1 qt

S: Solution E adds on the gallons first to get 9 gallons, then adds 2 quarts to make another gallon, and finally adds the one left over quart. → Solution F adds gallons first to get 9 gallons and then makes the next gallon to get 10 gallons 1 quart. → It's just like adding mixed numbers! Add the ones and then add the smaller units. → This time, Solution F just added like units to get 9 gallons 5 quarts and then took out the gallon from the 5 quarts.

Allow students to choose a method to solve and express the following sums with mixed units:

- 3 gallons 1 quart + 6 gallons 3 quarts
- 17 quarts 3 cups + 2 quarts 3 cups
- 4 gallons 7 pints + 10 gallons 7 pints

Problem 2: Subtract mixed units of capacity.

T: 4 cats – 3 cats is …?
S: 1 cat.
T: 4 fourths – 3 fourths is …?
S: 1 fourth.
T: (Write $1 - \frac{3}{4}$.) 1 minus 3 fourths is …?
S: 1 fourth.
T: (Directly below, write $8 - \frac{3}{4}$.) $8 - \frac{3}{4}$ is …?
S: $7\frac{1}{4}$.
T: Here are two different subtraction problems. Solve them with your partner, and then compare how they are similar to each other and to the problems you just solved with the fourths.

Problem 1
1 qt – 3 c

Problem 2
8 qt – 3 c
 /\
7 qt 4 c

S: 1 quart – 3 cups = 1 cup. 8 quarts – 3 cups = 7 quarts 1 cup. → You have to change 1 quart for 4 cups so you can subtract the cups. → It's like subtracting a fraction from a whole number, too. Actually, cups are like fourths in this problem! It takes 4 cups to make a quart just like it takes 4 fourths to make 1. So, you can change 1 quart to 4 cups just like you change 1 to 4 fourths.

Lesson 6: Solve problems involving mixed units of capacity.   85

Lesson 6   4•7

Have students solve the following:

- 1 gallon – 1 pint
- 8 gallons – 1 pint
- 1 quart – 2 cups
- 6 quarts – 2 cups

T: Here are two more subtraction problems. Solve them with your partner, and then compare them. How are they different? How are they the same?

Problem 3

8 qt 1 c − 3 c
 / \
7 qt   5 c

Problem 4

8 qt 1 c − 6 qt 3 c
 / \
7 qt   5 c

S: Problem 3 is a little trickier than Problem 2 because there is an extra cup. So, when you take 4 cups out of 8 quarts and 1 cup, you get 7 quarts and 5 cups because 4 cups + 1 cup is 5 cups. Now, you can subtract 3 cups. → In Problem 4, you have to subtract quarts, too, so just subtract like units. 7 quarts – 6 quarts is 1 quart. 5 cups – 3 cups is 2 cups. The answer is 1 quart 2 cups.

Have students solve the following:

- 9 gallons 2 quarts – 4 quarts
- 12 quarts 1 cup – 5 quarts 2 cups
- 6 gallons 3 pints – 2 gallons 7 pints

Note: Depending on how students are doing with the addition and subtraction of mixed capacity units, introduce compensation and counting up as exemplified below in the context of solving 8 quarts 1 cup – 6 quarts 3 cups. Solution A simply adds a cup to both the subtrahend and minuend (compensation). Solution B shows counting up from the subtrahend to the minuend.

Solution A

8 qt 1 c − 6 qt 3 c = 8 qt 2 c − 7 qt

Solution B

6 qt 3 c $\xrightarrow{+1c}$ 7 qt $\xrightarrow{1qt\,1c}$ 8 qt 1 c

Problem Set (10 minutes)

Students should do their personal best to complete the Problem Set within the allotted 10 minutes. For some classes, it may be appropriate to modify the assignment by specifying which problems they work on first. Some problems do not specify a method for solving. Students should solve these problems using the RDW approach used for Application Problems.

Student Debrief (12 minutes)

Lesson Objective: Solve problems involving mixed units of capacity.

The Student Debrief is intended to invite reflection and active processing of the total lesson experience.

Invite students to review their solutions for the Problem Set. They should check work by comparing answers with a partner before going over answers as a class. Look for misconceptions or misunderstandings that can be addressed in the Debrief. Guide students in a conversation to debrief the Problem Set and process the lesson.

Any combination of the questions below may be used to lead the discussion.

- What pattern did you notice between Problems 2(a) and 2(b)?
- When adding mixed units, we used two different strategies: adding like units and counting up with the arrow way. Was one strategy more effective? Did you prefer one strategy to another? Why?
- Explain to your partner how you solved Problem 4(a). Which strategy did you use for each of the ingredients?
- What was similar about working with gallons and quarts and quarts and cups?
- How is adding $5\frac{3}{4} + 7\frac{3}{4}$ like solving 5 gallons 3 quarts + 7 gallons 3 quarts?
- How is subtracting $5\frac{1}{8} - 2\frac{7}{8}$ like solving 5 gallons 1 pint − 2 gallons 7 pints?
- Compare using compensation to solve 81 − 29 or $8\frac{1}{4} - 2\frac{3}{4}$ to using compensation to solve 8 gallons 1 quart − 2 gallons 3 quarts.

Exit Ticket (3 minutes)

After the Student Debrief, instruct students to complete the Exit Ticket. A review of their work will help with assessing students' understanding of the concepts that were presented in today's lesson and planning more effectively for future lessons. The questions may be read aloud to the students.

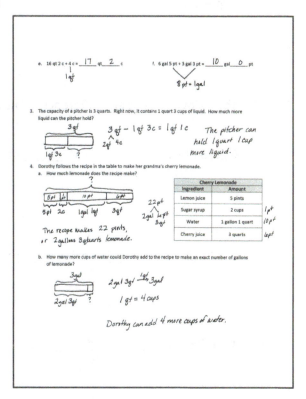

Lesson 6: Solve problems involving mixed units of capacity.

A STORY OF UNITS   Lesson 6 Problem Set   4•7

Name _____   Date _____

1. Determine the following sums and differences. Show your work.

   a. 3 qt + 1 qt = _____ gal   b. 2 gal 1 qt + 3 qt = _____ gal

   c. 1 gal – 1 qt = _____ qt   d. 5 gal – 1 qt = _____ gal _____ qt

   e. 2 c + 2 c = _____ qt   f. 1 qt 1 pt + 3 pt = _____ qt

   g. 2 qt – 3 pt = _____ pt   h. 5 qt – 3 c _____ qt _____ c

2. Find the following sums and differences. Show your work.

   a. 6 gal 3 qt + 3 qt = _____ gal _____ qt   b. 10 gal 3 qt + 3 gal 3 qt = _____ gal _____ qt

   c. 9 gal 1 pt – 2 pt = _____ gal _____ pt   d. 7 gal 1 pt – 2 gal 7 pt = _____ gal _____ pt

   e. 16 qt 2 c + 4 c = _____ qt _____ c   f. 6 gal 5 pt + 3 gal 3 pt = _____ gal _____ pt

Lesson 6:   Solve problems involving mixed units of capacity.

3. The capacity of a pitcher is 3 quarts. Right now, it contains 1 quart 3 cups of liquid. How much more liquid can the pitcher hold?

4. Dorothy follows the recipe in the table to make her grandma's cherry lemonade.

   a. How much lemonade does the recipe make?

Cherry Lemonade	
Ingredient	Amount
Lemon Juice	5 pints
Sugar Syrup	2 cups
Water	1 gallon 1 quart
Cherry Juice	3 quarts

   b. How many more cups of water could Dorothy add to the recipe to make an exact number of gallons of lemonade?

A STORY OF UNITS   Lesson 6 Exit Ticket   4•7

Name _____   Date _____

1. Find the following sums and differences. Show your work.

   a. 7 gal 2 qt + 3 gal 3 qt = _____ gal _____ qt

   b. 9 gal 1 qt – 5 gal 3 qt = _____ gal _____ qt

2. Jason poured 1 gallon 1 quart of water into an empty 2-gallon bucket. How much more water can be added to reach the bucket's 2-gallon capacity?

Name _____   Date _____

1. Determine the following sums and differences. Show your work.

   a. 5 qt + 3 qt = _____ gal

   b. 1 gal 2 qt + 2 qt = _____ gal

   c. 1 gal – 3 qt = _____ qt

   d. 3 gal – 2 qt = _____ gal _____ qt

   e. 1 c + 3 c = _____ qt

   f. 2 qt 3 c + 5 c = _____ qt

   g. 1 qt – 1 pt = _____ pt

   h. 6 qt – 5 pt = _____ qt _____ pt

2. Find the following sums and differences. Show your work.

   a. 4 gal 2 qt + 3 qt = _____ gal _____ qt

   b. 12 gal 2 qt + 5 gal 3 qt = _____ gal _____ qt

   c. 7 gal 2 pt – 3 pt = _____ gal _____ pt

   d. 11 gal 3 pt – 4 gal 6 pt = _____ gal _____ pt

   e. 12 qt 5 c + 6 c = _____ qt _____ c

   f. 8 gal 6 pt + 5 gal 4 pt = _____ gal _____ pt

Lesson 6: Solve problems involving mixed units of capacity.

3. The capacity of a bucket is 5 gallons. Right now, it contains 3 gallons 2 quarts of liquid. How much more liquid can the bucket hold?

4. Grace and Joyce follow the recipe in the table to make a homemade bubble solution.

   a. How much solution does the recipe make?

Homemade Bubble Solution	
Ingredient	Amount
Water	2 gallons 3 pints
Dish Soap	2 quarts 1 cup
Corn Syrup	2 cups

   b. How many more cups of solution would they need to fill a 4-gallon container?

A STORY OF UNITS

Lesson 7 4•7

Lesson 7

Objective: Solve problems involving mixed units of length.

Suggested Lesson Structure

■ Fluency Practice   (12 minutes)
■ Application Problem   (6 minutes)
■ Concept Development   (32 minutes)
■ Student Debrief   (10 minutes)
Total Time   **(60 minutes)**

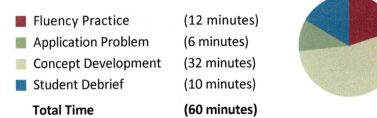

Fluency Practice (12 minutes)

- Grade 4 Core Fluency Differentiated Practice Sets **4.NBT.4**   (4 minutes)
- Add Mixed Numbers **4.NF.3c**   (4 minutes)
- Convert Length Units **4.MD.2**   (4 minutes)

Grade 4 Core Fluency Differentiated Practice Sets (4 minutes)

Materials: (S) Core Fluency Practice Sets (Lesson 2 Core Fluency Practice Sets)

Note: During Module 7, each day's Fluency Practice may include an opportunity for mastery of the addition and subtraction algorithm by means of the Core Fluency Practice Sets. The process is detailed and Practice Sets are provided in Lesson 2.

Add Mixed Numbers (4 minutes)

Materials: (S) Personal white board

Note: This fluency activity reviews Module 5's fraction work and anticipates today's lesson of adding mixed measurement units. Use choral or written responses as necessary.

T: 3 thirds + 5 thirds is how many thirds?
S: 8 thirds.
T: Express 8 thirds as ones and thirds.
S: 2 ones and 2 thirds.
T: 4 thirds + 9 thirds is how many thirds?
S: 13 thirds.

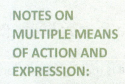

NOTES ON MULTIPLE MEANS OF ACTION AND EXPRESSION:

Enhance the learning experience by scaffolding the Add Mixed Numbers and Convert Length Units fluency activities for students working below grade level. Present a visual, such as a written form, or a model, such as a number bond, along with the questioning strategy. For example, "Express 8 thirds as ones and thirds."

Lesson 7:   Solve problems involving mixed units of capacity.

93

| A STORY OF UNITS | Lesson 7 | 4•7 |

T: Express 13 thirds as ones and thirds.

S: 4 ones and 1 third.

Continue with the following possible sequence: $\frac{7}{12} + \frac{8}{12}$, $\frac{5}{12} + \frac{9}{12}$, $\frac{14}{12} + \frac{3}{12}$, $\frac{9}{12} + \frac{9}{12}$.

Convert Length Units (4 minutes)

Materials: (S) Personal white board

Note: This fluency activity reviews Lesson 1 and anticipates today's work with length units. Use choral or written responses during the activity.

T: Express each number of yards and feet as feet.
T: 1 yard.
S: 3 feet.
T: 1 yard 2 feet.
S: 5 feet.
T: 4 yards 1 foot.
S: 13 feet.
T: 3 yards 2 feet.
S: 11 feet.
T: Express each number of feet as yards.
T: 3 feet is …?
S: 1 yard.
T: 6 feet is …?
S: 2 yards.
T: 9 feet is …?
S: 3 yards.

Repeat the process with feet and inches.

Application Problem (6 minutes)

Samantha is making punch for a class picnic. There are 26 students in her class. Samantha uses 1 gallon 2 quarts of orange juice, 3 quarts of lemonade, and 1 gallon 3 quarts of sparkling water. How much punch did Samantha make? Will there be enough for each student to have two 1-cup servings of punch?

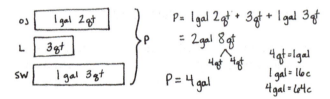

Note: This Application Problem links students' prior work with mixed units to today's work. Students review the skills of working with mixed units of capacity as a lead-in to today's Concept Development, where they work with mixed units of length.

A STORY OF UNITS　　　　　　　　　　　　　　　　　　　　　　　　　　　　　Lesson 7　4•7

Concept Development (32 minutes)

Materials: (S) Personal white board

Problem 1: Add mixed units of length.

- T: 8 months plus 7 months is how many months?
- S: 15 months.
- T: Express 15 months as years and months.
- S: 1 year 3 months.
- T: 8 twelfths plus 7 twelfths is how many twelfths?
- S: 15 twelfths.
- T: Express 15 twelfths as ones and twelfths.
- S: 1 one and 3 twelfths.
- T: 8 inches + 7 inches is how many inches?
- S: 15 inches.
- T: Express 15 inches as feet and inches.
- S: 1 foot 3 inches.
- T: Here are two different ways of adding 8 inches and 7 inches. Analyze them with your partner.

> **NOTES ON MULTIPLE MEANS OF ACTION AND EXPRESSION:**
>
> Today's lesson of partner analysis may be a welcome experience of autonomy and critical thinking for students working above grade level. Students working below grade level may benefit from more support through scaffolded questioning, visual models, and explicit instruction in adding and subtracting mixed units of measure.

Solution A

$8\text{in} \xrightarrow{+4\text{in}} 1\text{ft} \xrightarrow{+3\text{in}} 1\text{ft } 3\text{in}$

Solution B

$8\text{in} + 7\text{in} = 15\text{in} = 1\text{ft} + 3\text{in}$
           /\
        12in 3in

- S: Solution A makes 1 foot first by adding on 4 of the 7 inches and then adding on the other 3 inches. → Solution B adds the inches together to get 15 inches and then breaks the total into a foot and 3 inches. → Solution A is like adding 8 twelfths and 7 twelfths. You make one and then add on the extra 3 twelfths, the leftovers. → Solution B is like when we found the total number of months and then the number of years and extra months.
- T: Yes, we can either complete a foot and add on, or we can add all of the inches first and then break the total into feet and inches.

Allow students to choose a method to solve and express the following sums with mixed units of feet and inches or yards and feet:

- 11 inches + 9 inches
- 4 feet + 4 feet

Lesson 7:　Solve problems involving mixed units of capacity.　　95

A STORY OF UNITS   Lesson 7   4•7

T: Here are two different ways of adding 9 feet 8 inches + 7 inches. Analyze them with a partner. How are these problems like those we just solved?

Solution A
9 ft 8 in $\xrightarrow{+4in}$ 10 ft $\xrightarrow{+3in}$ 10 ft 3 in

Solution B
9 ft 8 in + 7 in = 9 ft 15 in = 10 ft 3 in
                                  /\
                                1 ft 3 in

S: Solution A breaks apart 7 inches as 4 inches and 3 inches and adds 4 inches first to make a foot.
→ Solution B adds like units. Since there is only 1 addend with feet as a unit, the solution combines the inches to get 15 inches, decomposes 15 inches, and then adds a foot to the 9 feet.

Allow students to choose a method to solve and express the following sums with mixed units:

- 4 feet 9 inches + 10 inches
- 6 yards 2 feet + 5 feet

T: Here are two different ways of adding 9 feet 8 inches + 6 feet 7 inches. Analyze them with a partner.

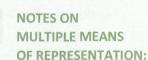

NOTES ON MULTIPLE MEANS OF REPRESENTATION:

If needed, allow English language learners to express their analysis in their native language. Provide sentence frames or starters to guide partner discussion, for example,

"Adding feet and inches is like adding cups, quarts, and gallons because …?"

or

"Adding mixed units is like adding mixed numbers because …?"

Solution A
9 ft 8 in $\xrightarrow{6ft}$ 15 ft 8 in $\xrightarrow{+7in}$ 16 ft 3 in

Solution B
9 ft 8 in + 6 ft 7 in = 15 ft 15 in = 16 ft 3 in
                                      /\
                                    1 ft 3 in

S: Both addends are mixed numbers, so Solution B just adds the like units right way. That seems easy to me. → In Solution A, we add each unit one at a time. First, we add feet to the first addend. Then, we add inches. The solution doesn't write out the breaking down of 7 inches as 4 inches and 3 inches. It's fair because we can do that with mental math now instead of writing everything out. → Adding mixed units is like adding two mixed numbers: Add the feet with the feet like we added the ones with the ones, and add the inches with the inches like we added the twelfths with the twelfths.

96   Lesson 7:   Solve problems involving mixed units of capacity.

A STORY OF UNITS　　　　　　　　　　　　　　　　　　　　　　　　　　　　　Lesson 7　4•7

T: Solve for 3 yards 2 feet + 2 yards 2 feet. Work with a partner. Try to solve it using a different solution method if you finish early.

Solution A

3 yd 2 ft + 2 yd 2 ft

3 yd 2 ft $\xrightarrow{2yd}$ 5 yd 2 ft $\xrightarrow{+1ft}$ 6 yd $\xrightarrow{+1ft}$ 6 yd 1 ft

Solution B

3 yd 2 ft + 2 yd 2 ft = 5 yd 4 ft = 6 yd 1 ft
　　　　　　　　　　　　　　　　　／＼
　　　　　　　　　　　　　　　1 yd　1 ft

S: We used the arrow way. First, we added the yards. Second, we broke apart the 2 feet to complete one more yard. Finally, we added the 1 foot that was left. Our answer is 6 yards 1 foot.

S: We added like units. Then, we broke 4 feet into 3 feet and 1 foot since 3 feet makes 1 more yard. Our answer is 6 yards 1 foot.

Allow students to choose a method to solve and express the following sums with mixed units:
4 feet 6 inches + 3 feet 11 inches and 6 yards 1 foot + 3 yards 2 feet.

Problem 2: Subtract mixed units of length.

T: $1 - \frac{9}{12}$ is ...?

S: 3 twelfths.

T: $7 - \frac{9}{12}$ is ...?

S: 6 and 3 twelfths.

T: 1 year – 9 months is how many months?

S: 3 months.

T: 7 years – 9 months is how many years and months?

S: 6 years 3 months.

T: (Write 1 foot – 9 inches and 7 feet – 9 inches.) Here are two different subtraction problems. Solve them with your partner, and compare how they are similar to each other and to the problems you just solved.

Problem 1

1 ft – 9 in = 12 in – 9 in = 3 in

Problem 2

7 ft – 9 in = 6 ft 3 in
／＼
6 ft　12 in

S: Feet and inches are like years and months. 1 year = 12 months just like 1 foot = 12 inches. We have to convert to inches to solve. In the first problem, there is only one foot. We convert it to inches. In the second problem, there are seven feet. We only need to convert one of the feet before subtracting.

EUREKA MATH　　Lesson 7:　Solve problems involving mixed units of capacity.　　97

T: (Write 7 feet 4 inches − 9 inches and 7 feet 4 inches − 5 feet 9 inches.) Solve with your partner by decomposing 1 foot into inches. How are these problems similar?

Problem 3
7 ft 4 in − 9 in = 6 ft 7 in
  / \
6 ft 16 in

Problem 4
7 ft 4 in − 5 ft 9 in = 1 ft 7 in
  / \
6 ft 16 in

S: Both problems decompose a foot into 12 inches so that we are able to subtract inches. → The difference is that in the second problem you have both feet and inches being subtracted. → You decompose 1 foot and then subtract feet from feet and inches from inches. → It's like when we subtract $7\frac{4}{12} - 5\frac{9}{12}$.

T: Now, try 7 yards 1 foot − 2 yards 2 feet.

S: We had to decompose 1 yard into 3 feet. That gave us 6 yards and 4 feet, and then we were able to subtract. 6 yards − 2 yards = 4 yards, and 4 feet − 2 feet = 2 feet. The difference is 4 yards 2 feet.

7 yd 1 ft − 2 yd 2 ft = 4 yd 2 ft
  / \
6 yd 4 ft

Have students solve 9 feet 2 inches − 4 feet 7 inches and 10 yards 1 foot − 7 yards 2 feet.

Depending on student progress with the addition and subtraction of mixed length units, possibly introduce compensation and counting up strategies as exemplified below in the context of solving 7 feet 4 inches − 5 feet 9 inches. Solution A shows simplification of the problem by adding 3 inches to both subtrahend and minuend. Solution B demonstrates counting up from the part being subtracted to the total.

Solution A
7 ft 4 in − 5 ft 9 in = 7 ft 7 in − 6 ft

Solution B
5 ft 9 in $\xrightarrow{+3in}$ 6 ft $\xrightarrow{+1ft\ 4in}$ 7 ft 4 in

Problem Set (10 minutes)

Students should do their personal best to complete the Problem Set within the allotted 10 minutes. For some classes, it may be appropriate to modify the assignment by specifying which problems they work on first. Some problems do not specify a method for solving. Students should solve these problems using the RDW approach used for Application Problems.

A STORY OF UNITS

Lesson 7   4•7

Student Debrief (10 minutes)

Lesson Objective: Solve problems involving mixed units of length.

The Student Debrief is intended to invite reflection and active processing of the total lesson experience.

Invite students to review their solutions for the Problem Set. They should check work by comparing answers with a partner before going over answers as a class. Look for misconceptions or misunderstandings that can be addressed in the Debrief. Guide students in a conversation to debrief the Problem Set and process the lesson.

Any combination of the questions below may be used to lead the discussion.

- How does Problem 2(a) relate to Problem 2(b)?
- Problems 3, 4, and 5 all seem to be very different problems. Explain how Problem 3 relates to Problem 5(a) and Problem 4 to Problem 5(b).
- Discuss with your partner how the strategies used today compare to the strategies used yesterday.
- Explain which strategy you like using best and why.
- How is solving 7 feet 4 inches – 5 feet 9 inches similar to solving $7\frac{4}{12} - 5\frac{9}{12}$?

Exit Ticket (3 minutes)

After the Student Debrief, instruct students to complete the Exit Ticket. A review of their work will help with assessing students' understanding of the concepts that were presented in today's lesson and planning more effectively for future lessons. The questions may be read aloud to the students.

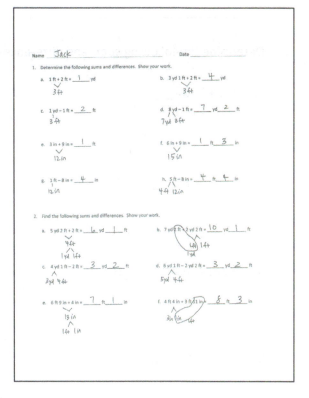

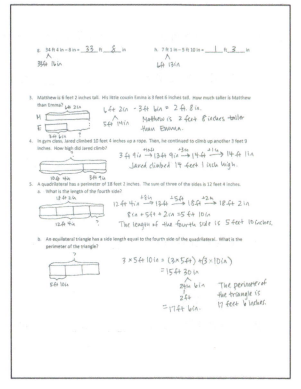

Lesson 7:   Solve problems involving mixed units of capacity.

A STORY OF UNITS

Lesson 7 Problem Set 4•7

Name _____ Date _____

1. Determine the following sums and differences. Show your work.

   a. 1 ft + 2 ft = _____ yd

   b. 3 yd 1 ft + 2 ft = _____ yd

   c. 1 yd – 1 ft = _____ ft

   d. 8 yd – 1 ft = _____ yd _____ ft

   e. 3 in + 9 in = _____ ft

   f. 6 in + 9 in = _____ ft _____ in

   g. 1 ft – 8 in = _____ in

   h. 5 ft – 8 in = _____ ft _____ in

2. Find the following sums and differences. Show your work.

   a. 5 yd 2 ft + 2 ft = _____ yd _____ ft

   b. 7 yd 2 ft + 2 yd 2 ft = _____ yd _____ ft

   c. 4 yd 1 ft – 2 ft = _____ yd _____ ft

   d. 6 yd 1 ft – 2 yd 2 ft = _____ yd _____ ft

   e. 6 ft 9 in + 4 in = _____ ft _____ in

   f. 4 ft 4 in + 3 ft 11 in = _____ ft _____ in

   g. 34 ft 4 in – 8 in = _____ ft _____ in

   h. 7 ft 1 in – 5 ft 10 in = _____ ft _____ in

Lesson 7: Solve problems involving mixed units of capacity.

3. Matthew is 6 feet 2 inches tall. His little cousin Emma is 3 feet 6 inches tall. How much taller is Matthew than Emma?

4. In gym class, Jared climbed 10 feet 4 inches up a rope. Then, he continued to climb up another 3 feet 9 inches. How high did Jared climb?

5. A quadrilateral has a perimeter of 18 feet 2 inches. The sum of three of the sides is 12 feet 4 inches.

   a. What is the length of the fourth side?

   b. An equilateral triangle has a side length equal to the fourth side of the quadrilateral. What is the perimeter of the triangle?

A STORY OF UNITS   Lesson 7 Exit Ticket   4•7

Name _____ Date _____

Determine the following sums and differences. Show your work.

1. 4 yd 1 ft + 2 ft _____ yd

2. 6 yd – 1 ft = _____ yd _____ ft

3. 4 yd 1 ft + 3 yd 2 ft = _____ yd

4. 8 yd 1 ft – 3 yd 2 ft = _____ yd _____ ft

Lesson 7: Solve problems involving mixed units of length.

A STORY OF UNITS   Lesson 7 Homework   4•7

Name _____   Date _____

1. Determine the following sums and differences. Show your work.

   a. 2 yd 2 ft + 1 ft = _____ yd   b. 2 yd – 1 ft = _____ yd _____ ft

   b. 2 ft + 2 ft = _____ yd _____ ft   d. 5 yd – 1 ft = _____ yd _____ ft

   e. 7 in + 5 in = _____ ft   f. 7 in + 7 in = _____ ft _____ in

   g. 1 ft – 2 in = _____ in   h. 2 ft – 6 in = _____ ft _____ in

2. Find the following sums and differences. Show your work.

   a. 4 yd 2 ft + 2 ft = _____ yd _____ ft   b. 6 yd 2 ft + 1 yd 1 ft = _____ yd _____ ft

   c. 5 yd 1 ft – 2 ft = _____ yd _____ ft   d. 7 yd 1 ft – 5 yd 2 ft = _____ yd _____ ft

   e. 7 ft 8 in + 5 in = _____ ft _____ in   f. 6 ft 5 in + 5 ft 9 in = _____ ft _____ in

   g. 32 ft 3 in – 7 in = _____ ft _____ in   h. 8 ft 2 in – 3 ft 11 in = _____ ft _____ in

Lesson 7:   Solve problems involving mixed units of length.

3. Laurie bought 9 feet 5 inches of blue ribbon. She also bought 6 feet 4 inches of green ribbon. How much ribbon did she buy altogether?

4. The length of the room is 11 feet 6 inches. The width of the room is 2 feet 9 inches shorter than the length. What is the width of the room?

5. Tim's bedroom is 12 feet 6 inches wide. The perimeter of the rectangular-shaped bedroom is 50 feet.

   a. What is the length of Tim's bedroom?

   b. How much longer is the length of Tim's room than the width?

A STORY OF UNITS   Lesson 8   4•7

Lesson 8

Objective: Solve problems involving mixed units of weight.

Suggested Lesson Structure

- Application Problem   (6 minutes)
- Fluency Practice   (12 minutes)
- Concept Development   (32 minutes)
- Student Debrief   (10 minutes)

  Total Time   **(60 minutes)**

Application Problem (6 minutes)

A sign next to the roller coaster says a person must be 54 inches tall to ride. At his last doctor's appointment, Hever was 4 feet 4 inches tall. He has grown 3 inches since then.

a. Is Hever tall enough to ride the roller coaster? By how many inches does he make or miss the minimum height?

b. Hever's father is 6 feet 3 inches tall. How much taller than the minimum height is his father?

$6ft\ 3in = 6ft + 3in = (6 \times 12in) + 3in = 72in + 3in = 75in$

$\begin{array}{r} 75 \\ -54 \\ \hline 21 \end{array}$ Hever's father is 21 inches taller than the minimum height.

Note: This Application Problem links students' prior work with mixed units to today's work. Students review the skills of working with mixed units of length to reinforce today's Concept Development, where they work with mixed units of weight.

Lesson 8:   Solve problems involving mixed units of weight.   105

A STORY OF UNITS   Lesson 8 4•7

Fluency Practice (12 minutes)

- Grade 4 Core Fluency Differentiated Practice Sets 4.NBT.4   (4 minutes)
- Add Mixed Numbers 4.MD.1   (4 minutes)
- Convert Weight Units 4.MD.1   (4 minutes)

Grade 4 Core Fluency Differentiated Practice Sets (4 minutes)

Materials: (S) Core Fluency Practice Sets (Lesson 2 Core Fluency Practice Sets)

Note: During Module 7, each day's Fluency Practice may include an opportunity for mastery of the addition and subtraction algorithm by means of the Core Fluency Practice Sets. The process is detailed and Practice Sets are provided in Lesson 2.

Add Mixed Numbers (4 minutes)

Materials: (S) Personal white board

Note: This fluency activity reviews Module 5's fraction work and anticipates today's lesson by adding mixed measurement units since sixteenths relate to pounds and ounces. Complete as a choral or white board activity.

- T: 8 sixteenths + 11 sixteenths is how many sixteenths?
- S: 19 sixteenths.
- T: Express 19 sixteenths as ones and sixteenths.
- S: 1 one and 3 sixteenths.
- T: 13 sixteenths + 8 sixteenths is how many sixteenths?
- S: 21 sixteenths.
- T: Express 21 sixteenths as ones and sixteenths.
- S: 1 one and 5 sixteenths.

Continue with the following possible sequence: $\frac{14}{16} + \frac{9}{16}$, $\frac{15}{16} + \frac{15}{16}$, $\frac{12}{16} + \frac{12}{16}$, $\frac{13}{16} + \frac{27}{16}$.

Convert Weight Units (4 minutes)

Materials: (S) Personal white board

Note: This fluency activity reviews Lesson 1 and anticipates today's work with weight units.

- T: Respond on your personal white board. Express each number of pounds and ounces as ounces.
- T: 1 pound?
- S: (Write 16 ounces.)
- T: 1 pound 3 ounces?
- S: (Write 19 ounces.)

---

106   Lesson 8:   Solve problems involving mixed units of weight.

A STORY OF UNITS   Lesson 8 4•7

T: 1 pound 1 ounce?
S: (Write 17 ounces.)
T: 2 pounds?
S: (Write 32 ounces.)
T: Express each number of ounces as pounds and ounces if possible.
T: 16 ounces is …?
S: (Write 1 pound.)
T: 32 ounces is …?
S: (Write 2 pounds.)

Repeat the process with 2 pounds 1 ounce, 2 pounds 11 ounces, 3 pound 15 ounces, and 3 pounds 6 ounces.

Concept Development (32 minutes)

Materials: (S) Personal white board

Note: The same lesson format may be followed from Lessons 6–7 if students need more guidance. This lesson invites students to share solution strategies on the assumption that they are ready to apply what they have learned in the previous two lessons to weight units.

NOTES ON MULTIPLE MEANS OF ENGAGEMENT:

If students working below grade level struggle with presenting strategies to solve, adjust the format of the lesson, offering guidance and practice until students become confident. Adding like units and decomposing the sum may be an approachable first strategy.

Challenge students working above grade level and others who may be ready to utilize, analyze, and discuss multiple strategies independently, within partnerships, or small groups.

Problem 1: Add mixed units of weight measure, and share alternate strategies.

T: (Write 4 lb 11 oz + 15 oz.) Solve the problem. Be prepared to share your solution strategy with a partner.
S: I completed a pound by adding 5 ounces and then added 10 more ounces. 5 pounds 10 ounces. (Solution A.)

Solution A
4 lb 11 oz —+5 oz→ 5 lb —+10 oz→ 5 lb 10 oz

S: I added a pound and then subtracted an ounce. (Solution B.)

Solution B
4 lb 11 oz —+1 lb→ 5 lb 11 oz —−1 oz→ 5 lb 10 oz

S: We can add up or add like units. That gave me 4 pounds 26 ounces, so I took out a pound from 26 ounces to find 5 pounds 10 ounces. (Solution C.)

Solution C
4 lb 11 oz + 15 oz = 4 lb 26 oz = 5 lb 10 oz
            16 oz 10 oz

Lesson 8:   Solve problems involving mixed units of weight.   107

A STORY OF UNITS   Lesson 8 4•7

Invite students to direct questions to their peers to understand their solution strategies.

T: (Display 24 lb 8 oz + 9 lb 13 oz.) Find this sum. Use the strategy you feel is most efficient.

S: I just added like units. (Solution A.)

Solution A
24 lb 8 oz + 9 lb 13 oz = 33 lb 21 oz = 34 lb 5 oz
                                    / \
                                 16 oz 5 oz

NOTES ON MULTIPLE MEANS OF ENGAGEMENT:

Empower English language learners to collaborate and communicate effectively to understand and explain solution strategies with sentence frames, sentence and question starters, and other scaffolds. Some possible question starters are given below:

- How did you solve?
- Why did you rename/compensate/count up/take from …?
- Can you explain why …?

S: I added the pounds first and then the ounces. (Solution B.)

MP.3

Solution B
24 lb 8 oz —+9lb→ 33 lb 8 oz —+13oz→ 33 lb 21 oz = 34 lb 5 oz
                                              / \
                                           1 lb 5 oz

S: I added 10 pounds and then subtracted 3 ounces since 9 lb 13 oz is 3 oz away from 10 lb. (Solution C.)

Solution C
24 lb 8 oz —+10 lb→ 34 lb 8 oz —-3 oz→ 34 lb 5 oz

Invite students to direct questions to their peers to understand their solution strategies.

Problem 2: Subtract units of weight measure when there are not enough smaller units.

T: (Display 6 lb 7 oz – 12 oz.) Solve the problem, and then you will have the opportunity to share your solution strategies with your peers.

Solution A
6 lb 7 oz – 12 oz = 5 lb 11 oz
      / \
   5 lb 23 oz

Solution B
6 lb 7 oz – 12 oz = 6 lb 11 oz – 1 lb = 5 lb 11 oz

Solution C
6 lb 7 oz —-7oz→ 6 lb —-5oz→ 5 lb 11 oz

Solution D
6 lb 7 oz – 12 oz = 5 lb 7 oz + 4 oz = 5 lb 11 oz
                         / \
                      5 lb 7 oz 16 oz

(I added 4 ounces to both the total and the part being subtracted.)

(I renamed a pound as 16 ounces and subtracted 12 ounces from 16 ounces.)

Lesson 8: Solve problems involving mixed units of weight.

A STORY OF UNITS

Lesson 8 4•7

The ability demonstrated on the previous page, to subtract a number of small units from a mixed number, is an essential skill. When students are ready, have them practice next with subtraction of a mixed number, such as 5 pounds 9 ounces – 2 pounds 14 ounces. This problem invites a variety of strategies, such as the following:

- Compensation: Students add 2 ounces to both minuend and subtrahend (since 14 ounces is 2 ounces away from 1 pound) to make the problem easier, as in 5 pounds 11 ounces – 3 pounds.
- Take from a pound: Students subtract the 2 pounds first and then subtract 14 ounces from 16 ounces.
- Rename a pound, and combine it with the ounces: Students rename 5 pounds 9 ounces as 4 pounds 25 ounces and subtract like units.
- Count up: Students count up 2 ounces to 3 pounds, add 2 pounds to get 5 pounds, and then add the remaining 9 ounces.

Problem Set (10 minutes)

Students should do their personal best to complete the Problem Set within the allotted 10 minutes. For some classes, it may be appropriate to modify the assignment by specifying which problems they work on first. Some problems do not specify a method for solving. Students should solve these problems using the RDW approach used for Application Problems.

Student Debrief (10 minutes)

Lesson Objective: Solve problems involving mixed units of weight.

The Student Debrief is intended to invite reflection and active processing of the total lesson experience.

Invite students to review their solutions for the Problem Set. They should check work by comparing answers with a partner before going over answers as a class. Look for misconceptions or misunderstandings that can be addressed in the Debrief. Guide students in a conversation to debrief the Problem Set and process the lesson.

Any combination of the questions below may be used to lead the discussion.

- What pattern did you notice between Problem 1(e) and Problem 1(f)?
- Explain to your partner how to solve Problem 1(g).
- For Problem 4(b), did you include the weight of the backpack as you calculated the answer? Does the weight of the backpack change the answer? Explain.

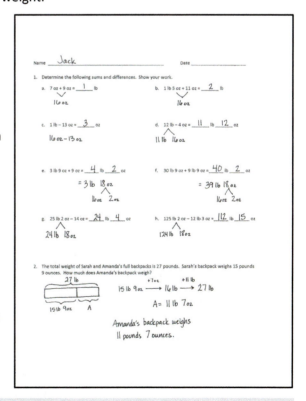

Lesson 8: Solve problems involving mixed units of weight.

- Explain how the work from Lessons 6, 7, and 8 are related.
- What makes one strategy for adding or subtracting mixed units more efficient than another?
- How is adding and subtracting weight measurement units like adding and subtracting mixed numbers? Length units? Capacity units?
- Notice that in the fluency activities we added sixteenths. Why do you think sixteenths were chosen as the unit in the fluency activities for this lesson?

Exit Ticket (3 minutes)

After the Student Debrief, instruct students to complete the Exit Ticket. A review of their work will help with assessing students' understanding of the concepts that were presented in today's lesson and planning more effectively for future lessons. The questions may be read aloud to the students.

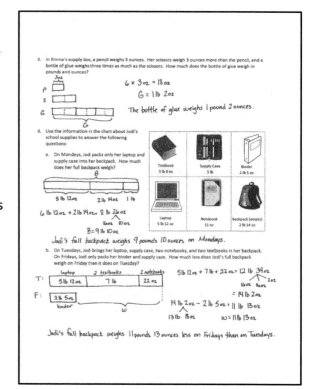

Name _____ Date _____

1. Determine the following sums and differences. Show your work.

   a. 7 oz + 9 oz = _____ lb

   b. 1 lb 5 oz + 11 oz = _____ lb

   c. 1 lb – 13 oz = _____ oz

   d. 12 lb – 4 oz = _____ lb _____ oz

   e. 3 lb 9 oz + 9 oz = _____ lb _____ oz

   f. 30 lb 9 oz + 9 lb 9 oz _____ lb _____ oz

   g. 25 lb 2 oz – 14 oz = _____ lb _____ oz

   h. 125 lb 2 oz – 12 lb 3 oz = _____ lb _____ oz

2. The total weight of Sarah and Amanda's full backpacks is 27 pounds. Sarah's backpack weighs 15 pounds 9 ounces. How much does Amanda's backpack weigh?

3. In Emma's supply box, a pencil weighs 3 ounces. Her scissors weigh 3 ounces more than the pencil, and a bottle of glue weighs three times as much as the scissors. How much does the bottle of glue weigh in pounds and ounces?

4. Use the information in the chart about Jodi's school supplies to answer the following questions:

   a. On Mondays, Jodi packs only her laptop and supply case into her backpack. How much does her full backpack weigh?

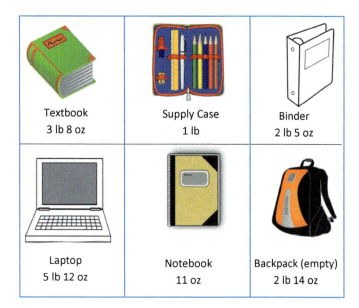

Textbook
3 lb 8 oz

Supply Case
1 lb

Binder
2 lb 5 oz

Laptop
5 lb 12 oz

Notebook
11 oz

Backpack (empty)
2 lb 14 oz

   b. On Tuesdays, Jodi brings her laptop, supply case, two notebooks, and two textbooks in her backpack. On Fridays, Jodi only packs her binder and supply case. How much less does Jodi's full backpack weigh on Friday than it does on Tuesday?

A STORY OF UNITS　　　　　　　　　　　　　　　　　　　Lesson 8 Exit Ticket　4•7

Name _____ Date _____

Determine the following sums and differences. Show your work.

1. 4 lb 6 oz + 10 oz = _____ lb _____ oz

2. 12 lb 4 oz + 3 lb 14 oz = _____ lb _____ oz

3. 5 lb 4 oz − 12 oz = _____ lb _____ oz

4. 20 lb 5 oz − 13 lb 7 oz = _____ lb _____ oz

Lesson 8:　Solve problems involving mixed units of weight.

A STORY OF UNITS

Lesson 8 Homework 4•7

Name _____ Date _____

1. Determine the following sums and differences. Show your work.

   a. 11 oz + 5 oz = _____ lb

   b. 1 lb 7 oz + 9 oz = _____ lb

   c. 1 lb – 11 oz = _____ oz

   d. 12 lb – 8 oz = _____ lb _____ oz

   e. 5 lb 8 oz + 9 oz = _____ lb _____ oz

   f. 21 lb 8 oz + 6 lb 9 oz = _____ lb _____ oz

   g. 23 lb 1 oz – 15 oz = _____ lb _____ oz

   h. 89 lb 2 oz – 16 lb 4 oz = _____ lb _____ oz

2. When David took his dog, Rocky, to the vet in December, Rocky weighed 29 pounds 9 ounces. When he took Rocky back to the vet in March, Rocky weighed 34 pounds 4 ounces. How much weight did Rocky gain?

3. Bianca had 6 identical jars of bubble bath. She put them all in a bag that weighed 2 ounces. The total weight of the bag filled with the six jars was 1 pound 4 ounces. How much did each jar weigh?

Lesson 8: Solve problems involving mixed units of weight.

4. Use the information in the chart about Melissa's school supplies to answer the following questions:

   a. On Wednesdays, Melissa packs only two notebooks and a binder into her backpack. How much does her full backpack weigh on Wednesdays?

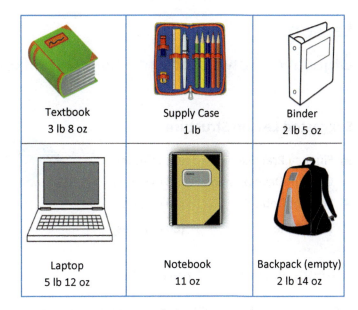

   b. On Thursdays, Melissa puts her laptop, supply case, two textbooks, and a notebook in her backpack. How much does her full backpack weigh on Thursdays?

   c. How much more does the backpack weigh with 3 textbooks and a notebook than it does with just 1 textbook and the supply case?

Lesson 9

Objective: Solve problems involving mixed units of time.

Suggested Lesson Structure

- ■ Fluency Practice   (12 minutes)
- □ Concept Development   (38 minutes)
- ■ Student Debrief   (10 minutes)
- **Total Time**   **(60 minutes)**

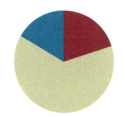

Fluency Practice (12 minutes)

- Grade 4 Core Fluency Differentiated Practice Sets **4.NBT.4**   (4 minutes)
- Add Mixed Numbers **4.NF.4**   (4 minutes)
- Convert Time Units **4.MD.1**   (4 minutes)

Grade 4 Core Fluency Differentiated Practice Sets (4 minutes)

Materials: (S) Core Fluency Practice Sets (Lesson 2 Core Fluency Practice Sets)

Note: During Module 7, each day's Fluency Practice may include an opportunity for mastery of the addition and subtraction algorithm by means of the Core Fluency Practice Sets. The process is detailed and Practice Sets are provided in Lesson 2.

Add Mixed Numbers (4 minutes)

Materials: (S) Personal white board

Note: This fluency activity reviews Module 5's fraction work and anticipates today's lesson of adding mixed measurement units, specifically twenty-fourths and sixtieths, to prepare for work with the hours in a day, the seconds in a minute, and the minutes in an hour. Complete as a choral or white board activity.

- T: 10 twenty-fourths + 17 twenty-fourths is how many twenty-fourths?
- S: 27 twenty-fourths.
- T: Express 27 twenty-fourths as ones and twenty-fourths.
- S: 1 one and 3 twenty-fourths.
- T: 20 twenty-fourths + 20 twenty-fourths is how many twenty-fourths?
- S: 40 twenty-fourths.
- T: Express 40 twenty-fourths as ones and twenty-fourths.
- S: 1 one and 16 twenty-fourths.

Continue using the following possible sequence: $\frac{50}{60}+\frac{20}{60}, \frac{15}{60}+\frac{45}{60}, \frac{30}{60}+\frac{45}{60}, \frac{45}{60}+\frac{45}{60}$.

A STORY OF UNITS　　　　　　　　　　　　　　　　　　　　　　　　　　　　Lesson 9　4•7

Convert Time Units (4 minutes)

Materials: (S) Personal white board

Note: This fluency activity reviews Lesson 3 and anticipates the lesson's work with time units. Complete as a choral or white board activity.

- T: Express each number of days and hours as hours.
- T: 1 day.
- S: 24 hours.
- T: 1 day 3 hours.
- S: 27 hours.
- T: 1 day 1 hour.
- S: 25 hours.
- T: 2 days.
- S: 48 hours.
- T: Express each number of hours as days and hours.
- T: 24 hours is …?
- S: 1 day.
- T: 48 hours is …?
- S: 2 days.
- T: 72 hours is …?
- S: 3 days.

Repeat the same process with hours and minutes.

Concept Development (38 minutes)

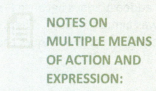

NOTES ON MULTIPLE MEANS OF ACTION AND EXPRESSION:

In keeping with the previous lessons of exploration, analysis, and autonomy, today's lesson may be a welcome experience of independence and critical thinking for students working above grade level. Students working below grade level may benefit from more support through scaffolded questioning, visual models, and explicit instruction as to how to add and subtract mixed units of measure.

Materials: (S) Personal white board

Problem 1: Add mixed units of time, and share alternate strategies.

Note: The same lesson format may be followed from Lessons 6–8 if so desired. This lesson invites students to share solution strategies on the assumption that they are ready to apply what they have learned in the previous three lessons to time units.

- T: (Display 2 hr 45 min + 50 min.) Solve this problem, and be prepared to share your solution strategy.
- S: I decomposed 50 minutes to complete an hour and added on the extra minutes. (Solution A.)
- S: I added an hour first and subtracted 10 minutes from my answer because 50 minutes is 10 minutes less than 1 hour. (Solution B.)

Lesson 9:　　Solve problems involving mixed units of time.　　　　　117

S: I added the minutes and then took out 60 minutes from the total number of minutes. (Solution C.)

2 hr 45 min + 50 min

Solution A
2 hr 45 min —+15 min→ 3 hr —+35min→ 3 hr 35 min

Solution B
2 hr 45 min —+1hr→ 3 hr 45 min —−10min→ 3 hr 35 min

Solution C
2 hr 45 min + 50 min = 2 hr 95 min = 3 hr 35 min
            60 min 35 min

Invite students to direct questions to their peers to understand their solution strategies. If students seem ready to move on to the addition of a mixed unit, continue into the next set. If not, give additional practice with problems such as 4 days 16 hours + 8 hours and 8 minutes 47 seconds + 36 seconds.

T: (Display 3 days 12 hours + 9 days 20 hours.) Find the sum. Use the strategy you feel is most efficient.
S: I added the days first. Next, I completed a day by adding on 12 hours. Finally, I knew there were 8 more hours to add on. (Solution A.)
S: I added like units and then took out a day from the total number of hours. (Solution B.)
S: I added 10 days because I realized that 9 days 20 hours was almost 10 days. Then, I subtracted 4 hours to make up for the 4 hours I added on. (Solution C.)

NOTES ON MULTIPLE MEANS OF ACTION AND EXPRESSION:

Some learners may benefit from a modeling of Solution C as a think aloud. Learners may benefit from understanding the circumstances in which this strategy is beneficial to use and when it is not.

Solution A
3 days 12 hr —+9 days→ 12 days 12 hr —+12 hrs→ 13 days —+8 hrs→ 13 days 8 hr.

Solution B
3 days 12 hrs + 9 days 20 hr = 12 days 32 hr = 13 days 8 hr.
                                    1 day 8 hr

Solution C
3 days 12 hrs —+10 days→ 13 days 12 hr —−4 hr→ 13 days 8 hr.

Let students continue to practice adding mixed units of time using the following: 12 hr 45 min + 3 hr 45 min, 19 min 15 sec + 6 min 58 sec, 2 days 19 hours + 6 days 13 hours, and 24 min 10 sec + 9 min 53 sec.

Problem 2: Subtract units of time when there are not enough smaller units.

T: (Display 7 hr 15 min – 38 min.) What is different about this problem? Use what you know to solve.

S: There are not enough minutes to subtract. I subtracted 15 minutes to get to 7 hours and then subtracted 23 more minutes to get to 6 hours and 37 minutes. (Solution A.)

S: I renamed an hour as 60 minutes to get 6 hours and 75 minutes and then just subtracted 38 minutes from 75 minutes. (Solution B.)

S: I renamed 7 hr 15 min to 6 hr 15 min + 60 min. Next, I subtracted 38 min from 60 min and got 22 min. Finally, I added the remaining hours and minutes to make 6 hr 37 min. (Solution C.)

S: I added 22 minutes to both the total and the part being subtracted to make it easy. Just subtract an hour. (Solution D.)

Solution A
7 hr 15 min $\xrightarrow{-15\text{min}}$ 7 hr $\xrightarrow{-23\text{min}}$ 6 hr 37 min

Solution B
7 hr 15 min − 38 min = 6 hr 37 min
       / \
   6 hr   75 min

Solution C
7 hr 15 min − 38 min = 6 hr 15 min + 22 min = 6 hr 37 min
       / \
  6 hr 15min 60 min

Solution D
7 hr 15 min − 38 min = 7 hr 37 min − 1 hr = 6 hr 37 min

Invite students to direct questions to their peers to understand their solution strategies. If students seem ready to move on to the subtraction of a mixed unit, continue into the next set. If not, give additional practice with problems such as 11 days 10 hours – 16 hours or 8 minutes 12 seconds – 36 seconds.

T: (Display 25 min 8 sec – 12 min 46 sec.) Use the strategy you feel is most efficient. Find the difference.

25 min 8 sec − 12 min 46 sec.

Solution A
25 min 8 sec $\xrightarrow{-12\text{min}}$ 13 min 8 sec $\xrightarrow{-8\text{sec}}$ 13 min $\xrightarrow{-38\text{sec}}$ 12 min 22 sec

Solution B
25 min 8 sec − 12 min 46 sec. = 12 min 22 sec
    / \
 24 min 68 sec

Solution C
25 min 8 sec − 12 min 46 sec = 25 min 22 sec − 13 min = 12 min 22 sec.

S: I subtracted 12 minutes first. Next, I subtracted 8 seconds to get to 13 minutes and then took away the rest of the seconds. (Solution A.)

Lesson 9: Solve problems involving mixed units of time.

A STORY OF UNITS   Lesson 9   4•7

S: I renamed 25 minutes 8 seconds as 24 minutes 68 seconds and then just subtracted minutes from minutes and seconds from seconds. (Solution B.)

S: I added 14 seconds to both numbers in order to just subtract 13 minutes. (Solution C.)

Let students practice finding the difference between mixed units of time using the following: 60 min 2 sec – 12 minutes 4 sec, 16 hr 10 min – 15 hr 15 min, and 17 days 3 hours – 10 days 14 hours.

Problem Set (10 minutes)

Students should do their personal best to complete the Problem Set within the allotted 10 minutes. For some classes, it may be appropriate to modify the assignment by specifying which problems they work on first. Some problems do not specify a method for solving. Students should solve these problems using the RDW approach used for Application Problems.

Student Debrief (10 minutes)

Lesson Objective: Solve problems involving mixed units of time.

The Student Debrief is intended to invite reflection and active processing of the total lesson experience.

Invite students to review their solutions for the Problem Set. They should check work by comparing answers with a partner before going over answers as a class. Look for misconceptions or misunderstandings that can be addressed in the Debrief. Guide students in a conversation to debrief the Problem Set and process the lesson.

Any combination of the questions below may be used to lead the discussion.

- How was solving Problem 2(a) similar to solving 2(b)? How was it different?
- Many of you solved Problem 4(b) by adding the two movie times together with the 30 extra minutes and then subtracting that time from 5 hours. Talk with your partner about how to use your answer from Problem 4(a) to help solve 4(b).
- How is solving 3 days 12 hours + 9 days 20 hours like solving $3\frac{12}{24} + 9\frac{20}{24}$?
- How is subtracting 25 min 8 sec – 12 min 46 sec like solving $25\frac{8}{60} - 12\frac{46}{60}$?
- How is solving 3 days 12 hours + 9 days 20 hours like solving 3 pounds 12 ounces + 9 pounds 8 ounces? How is it different?

- How did our fluency activities prepare us for our lesson?

Exit Ticket (3 minutes)

After the Student Debrief, instruct students to complete the Exit Ticket. A review of their work will help with assessing students' understanding of the concepts that were presented in today's lesson and planning more effectively for future lessons. The questions may be read aloud to the students.

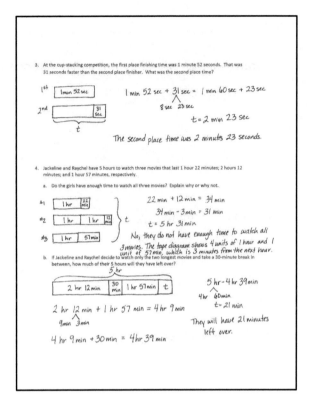

A STORY OF UNITS   Lesson 9 Problem Set   4•7

Name _____   Date _____

1. Determine the following sums and differences. Show your work.

   a. 23 min + 37 min = _____ hr   b. 1 hr 11 min + 49 min = _____ hr

   c. 1 hr – 12 min = _____ min   d. 4 hr – 12 min = _____ hr _____ min

   e. 22 sec + 38 sec = _____ min   f. 3 min – 45 sec = _____ min _____ sec

2. Find the following sums and differences. Show your work.

   a. 3 hr 45 min + 25 min = _____ hr _____ min   b. 2 hr 45 min + 6 hr 25 min = _____ hr _____ min

   c. 3 hr 7 min – 42 min = _____ hr _____ min   d. 5 hr 7 min – 2 hr 13 min = _____ hr _____ min

   e. 5 min 40 sec + 27 sec = _____ min _____ sec f. 22 min 48 sec – 5 min 58 sec = _____ min _____ sec

Lesson 9: Solve problems involving mixed units of time.

3. At the cup-stacking competition, the first place finishing time was 1 minute 52 seconds. That was 31 seconds faster than the second place finisher. What was the second place time?

4. Jackeline and Raychel have 5 hours to watch three movies that last 1 hour 22 minutes, 2 hours 12 minutes, and 1 hour 57 minutes, respectively.

   a. Do the girls have enough time to watch all three movies? Explain why or why not.

   b. If Jackeline and Raychel decide to watch only the two longest movies and take a 30-minute break in between, how much of their 5 hours will they have left over?

A STORY OF UNITS — Lesson 9 Exit Ticket 4•7

Name _____ Date _____

Find the following sums and differences. Show your work.

1. 2 hr 25 min + 25 min = _____ hr _____ min

2. 4 hr 45 min + 2 hr 35 min = _____ hr _____ min

3. 11 hr 6 min – 32 min = _____ hr _____ min

4. 8 hr 9 min – 6 hr 42 min = _____ hr _____ min

Lesson 9: Solve problems involving mixed units of time.

A STORY OF UNITS

Lesson 9 Homework 4•7

Name _____ Date _____

1. Determine the following sums and differences. Show your work.

   a. 41 min + 19 min = _____ hr

   b. 2 hr 21 min + 39 min = _____ hr

   c. 1 hr − 33 min = _____ min

   d. 3 hr − 33 min = _____ hr _____ min

   e. 31 sec + 29 sec = _____ min

   f. 5 min − 15 sec = _____ min _____ sec

2. Find the following sums and differences. Show your work.

   a. 5 hr 30 min + 35 min = _____ hr _____ min

   b. 3 hr 15 min + 5 hr 55 min = _____ hr _____ min

   c. 4 hr 4 min − 38 min = _____ hr _____ min

   d. 7 hr 3 min − 4 hr 25 min = _____ hr _____ min

   e. 3 min 20 sec + 49 sec = _____ min _____ sec

   f. 22 min 37 sec − 5 min 58 sec = _____ min _____ sec

Lesson 9: Solve problems involving mixed units of time.

3. It took 5 minutes 34 seconds for Melissa's oven to preheat to 350 degrees. That was 27 seconds slower than it took Ryan's oven to preheat to the same temperature. How long did it take Ryan's oven to preheat?

4. Joanna read three books. Her goal was to finish all three books in a total of 7 hours. She completed them, respectively, in 2 hours 37 minutes, 3 hours 9 minutes, and 1 hour 51 minutes.

   a. Did Joanna meet her goal? Write a statement to explain why or why not.

   b. Joanna completed the two shortest books in one evening. How long did she spend reading that evening? How long, with her goal in mind, did that leave her to read the third book?

Lesson 10

Objective: Solve multi-step measurement word problems.

Suggested Lesson Structure

- ■ Fluency Practice   (12 minutes)
- ■ Concept Development (35 minutes)
- ■ Student Debrief   (13 minutes)
- **Total Time**   **(60 minutes)**

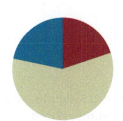

Fluency Practice (12 minutes)

- Grade 4 Core Fluency Differentiated Practice Sets **4.NBT.4**   (4 minutes)
- Add Mixed Numbers **4.MD.2**   (4 minutes)
- Convert Capacity and Length Units **4.MD.1**   (4 minutes)

Grade 4 Core Fluency Differentiated Practice Sets (4 minutes)

Materials: (S) Core Fluency Practice Sets (Lesson 2 Core Fluency Practice Sets)

Note: During Module 7, each day's Fluency Practice may include an opportunity for mastery of the addition and subtraction algorithm by means of the Core Fluency Practice Sets. The process is detailed and Practice Sets are provided in Lesson 2.

Add Mixed Numbers (4 minutes)

Materials: (S) Personal white board

Note: This fluency activity reviews Module 5's fraction work and anticipates today's lesson of adding mixed measurement units. Direct students to respond chorally to the questions or use a written response on their personal white boards, depending on which is most effective for them.

T: 3 fourths + 3 fourths is how many fourths?
S: 6 fourths.
T: Express 6 fourths as ones and fourths.
S: 1 one and 2 fourths.
T: 3 quarts + 3 quarts is how many quarts?
S: 6 quarts.
T: Express 6 quarts as gallons and quarts. Draw a number bond to pull out 4 quarts.
S: 1 gallon 2 quarts.

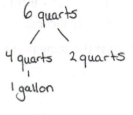

A STORY OF UNITS   Lesson 10   4•7

T: 7 twelfths + 7 twelfths is how many twelfths?
S: 14 twelfths.
T: Express 14 twelfths as ones and twelfths.
S: 1 one and 2 twelfths.
T: 7 inches + 7 inches is how many inches?
S: 14 inches.
T: Express 14 inches as feet and inches. Draw a number bond to pull out 12 inches.
S: 1 foot 2 inches.

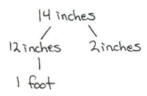

Continue with the following possible sequence: 6 eighths + 6 eighths related to 6 pints + 6 pints, and 11 sixteenths + 11 sixteenths related to 11 ounces + 11 ounces.

Convert Capacity and Length Units (4 minutes)

Materials: (S) Personal white board

Note: This fluency activity reviews Lessons 1–2 and anticipates today's work with capacity and length units. Direct students to respond chorally to the questions at a signal or to use written responses on their personal white boards, depending on which is most effective for them.

T: Express each number of quarts and cups as cups.
T: 1 quart.
S: 4 cups.
T: 1 quart 2 cups.
S: 6 cups.
T: Express each number of feet and inches as inches.
T: 1 foot 1 inch.
S: 13 inches.
T: 2 quarts 3 cups.
S: 11 cups.
T: 3 feet 7 inches.
S: 43 inches.

> **NOTES ON MULTIPLE MEANS OF REPRESENTATION:**
>
> To clarify the Convert Capacity and Length Units fluency activity directions for English language learners and others, give an example demonstrating the anticipated response.

Repeat the same process with gallons and pints and then yards and feet.

Lesson 10: Solve multi-step measurement word problems.

Lesson 10 4•7

Concept Development (35 minutes)

Materials: (S) Problem Set

Note: The sample solutions for each problem are examples of the type of thinking that students might use in solving each problem. The solutions are not inclusive of all possible strategies. Encourage and challenge students to explain the strategies that they use.

Suggested Delivery of Instruction for Solving Lesson 10's Word Problems

For Problems 1–4 below, students may work in pairs to solve each of the problems using the RDW approach to problem solving.

1. Model the problem.

Select two pairs of students who can successfully model the problem to work at the board while the other students work independently or in pairs at their seats. Review the following questions before beginning the first problem.

- Can you draw something?
- What can you draw?
- What conclusions can you make from your drawing?

As students work, circulate. Reiterate the questions above. After two minutes, have the two pairs of students share only their labeled diagrams. For about one minute, have the demonstrating students receive and respond to feedback and questions from their peers.

> **NOTES ON MULTIPLE MEANS OF ENGAGEMENT:**
>
> Communicate clear expectations for modeling that allow all students to understand what it takes to become a demonstrating student. Offering a rubric and scaffolds by which students can set and achieve goals may give everyone a fair chance to succeed. Demonstrating students may use translators, interpreters, or sentence frames to present and respond to feedback.

2. Calculate to solve and write a statement.

Allow students two minutes to complete work on the problem, sharing their work and thinking with a peer. Have students write their equations and statements of the answer.

3. Assess the solution.

Give students one to two minutes to assess the solutions presented by their peers on the board, comparing the solutions to their own work. Highlight alternative methods to reach the correct solution.

Lesson 10: Solve multi-step measurement word problems.

Problem 1

Paula's time swimming in the Ironman Triathlon was 1 hour 25 minutes. Her time biking was 5 hours longer than her swimming time. She ran for 4 hours 50 minutes. How long did it take her to complete all three parts of the race?

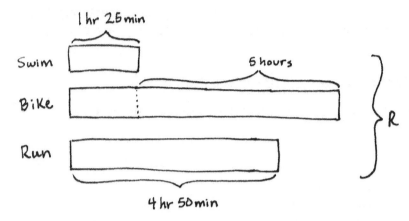

It took Paula 12 hours 40 minutes to complete the whole race.

Solution A
R = 1 hr 25 min + 6 hr 25 min + 4 hr 50 min
  = 11 hr 100 min
       ∧
    60 min 40 min
R = 12 hr 40 min

Solution B

1 hr 25 min →(+1hr 25min)→ 2 hr 50 min →(+5hr)→ 7 hr 50 min

7 hr 50 min →(+4hr)→ 11 hr 50 min →(+10min)→ 12 hr →(+40min)→ 12 hr 40 min

R = 12h 40min

This problem could be solved, as in Solution A, by adding like units. Students may also, as in Solution B, solve by adding up. First, the student adds the 2 equal units of 1 hour 25 minutes and then adds the additional 5 hours. Then, the student adds the remaining 4 hours and 50 minutes, decomposing 50 minutes into 10 minutes and 40 minutes as to complete the whole, the next hour.

A STORY OF UNITS

Lesson 10 4•7

Problem 2

Nolan put 7 gallons 3 quarts of gas into his car on Monday and twice as much on Saturday. What was the total amount of gas put into the car on both days?

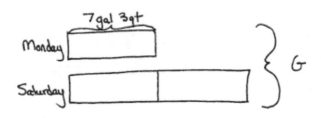

Solution A

1 gallon = 4 quarts
7 gallons = 28 quarts
28 quarts + 3 quarts = 31 quarts
1 unit = 31 quarts
3 units = 93 quarts
G = 93 quarts

Nolan put 93 quarts of gas into his car.

Solution B

3 × 7 gallons = 21 gallons
3 × 3 quarts = 9 quarts = 2 gallons 1 quart
          / | \
        4qt 4qt 1qt
21 gallons + 2 gallons 1 quart = 23 gallons 1 quart
G = 23 gallons 1 quart

Nolan put 23 gallons 1 quart of gas into his car.

Solution C

8 gal × 3 = 24 gal

24 gal − 3 qt = 23 gal 1 qt
   ^
 23 gal 4qt

MP.2

Once students see the relationship between the amount of gas added on Monday and Saturday, they can use different strategies to figure out how much gas was put in the car. The amount of gas can be converted into quarts, as modeled in Solution A, or the student may work with the mixed units to get 23 gallons 1 quart of gas, as shown in Solution B. Solution C shows an alternative method of rounding the gas for each unit to 8 gallons, finding that about 24 gallons of gas was put into Nolan's car. Each unit was rounded up by 1 quart, so then 3 quarts—or 1 quart for each unit—is subtracted from 24 gallons.

Lesson 10: Solve multi-step measurement word problems.

Problem 3

One pumpkin weighs 7 pounds 12 ounces. A second pumpkin weighs 10 pounds 4 ounces. A third pumpkin weighs 2 pounds 9 ounces more than the second pumpkin. What is the total weight of all 3 pumpkins?

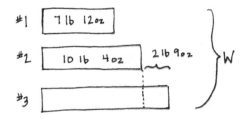

The total weight of all 3 pumpkins is 30 pounds 13 ounces.

Solution A

10 lb 4 oz —+2lb 9oz→ 12 lb 13 oz

7 lb 12 oz —+10lb 4oz→ 18 lb —+12lb 13oz→ 30 lb 13 oz

W = 30 lb 13 oz

Solution B

10 lb 4 oz + 2 lb 9 oz = 12 lb 13 oz

W = 7 lb 12 oz + 10 lb 4 oz + 12 lb 13 oz

= 29 lb 29 oz
         ∧
      16 oz 13 oz

W = 30 lb 13 oz

Solution C

2 × (10 lb 4 oz)

= 2 × (10 lb + 4 oz)

= (2 × 10 lb) + (2 × 4 oz)

= 20 lb + 8 oz

20 lb 8 oz + 2 lb 9 oz = 22 lb 17 oz
                              ∧
                           16 oz 1 oz

= 23 lb 1 oz

23 lb 1 oz + 7 lb 12 oz = 30 lb 13 oz

W = 30 lb 13 oz

Solution A models the arrow way of adding up. First, the weight of the third pumpkin is determined. Next, the three weights are added together to find their total weight. Solution B uses mixed unit addition, first finding the weight of the third pumpkin and then adding all three weights together. A number bond shows how 1 pound can be taken out of 29 ounces, just as 1 whole can be taken out of 5 fourths. Solution C models using multiplication to find the weight of the full unit of the second pumpkin and the partial unit of the third pumpkin. Then, the additional weight of the third pumpkin and the weight of the first pumpkin are added on. All three solutions shown are computed in mixed units because converting all weights to ounces and then finding their sum would be an inefficient, but possible, strategy.

Problem 4

Mr. Lane is 6 feet 4 inches tall. His daughter, Mary, is 3 feet 8 inches shorter than her father. His son is 9 inches taller than Mary. How many inches taller is Mr. Lane than his son?

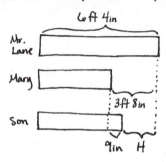

H = 2 ft 11 in
= 24 in + 11 in
= 35 in

Mr. Lane is 35 inches taller than his son.

Solution A
3 ft 8 in − 9 in
   ∧
2 ft 12 in
2 ft 20 in − 9 in = 2 ft 11 in

Solution B
6 ft 4 in − 3 ft 8 in
   ∧
5 ft 12 in
5 ft 16 in − 3 ft 8 in = 2 ft 8 in (Mary)
2 ft 8 in + 9 in = 2 ft 17 in
                        ∧
                      12 in 5 in
= 3 ft 5 in (Son)
6 ft 4 in − 3 ft 5 in = 2 ft 11 in
   ∧
5 ft 12 in

As in Solution A, students may notice from the tape diagrams that they don't need to find Mary's height or the son's height to solve this problem. They can subtract the 9 inches from the 3 feet 8 inches to see how much taller Mr. Lane is than his son. As shown in Solution B, students can use the given information to find Mary's height and then add 9 inches to find the son's height. The son's height can be subtracted from Mr. Lane's height to find the difference, and then the difference can be converted to inches to find the solution. Breaking out a foot to subtract the inches makes the subtraction process easier.

Problem Set

Please note that the Problem Set is completed as part of the Concept Development for this lesson.

Student Debrief (13 minutes)

Lesson Objective: Solve multi-step measurement word problems.

The Student Debrief is intended to invite reflection and active processing of the total lesson experience.

Invite students to review their solutions for the Problem Set. They should check work by comparing answers with a partner before going over answers as a class. Look for misconceptions or misunderstandings that can be addressed in the Debrief. Guide students in a conversation to debrief the Problem Set and process the lesson.

A STORY OF UNITS

Lesson 10 4•7

Any combination of the questions below may be used to lead the discussion.

- Look at Problem 2. Discuss with your partner which of your solutions is more efficient.
- Is it more efficient to add or multiply for Problem 2? How does that choice affect the units of the solution?
- Explain to your partner how you solved Problem 3. If you used different strategies, discuss how you arrived at the same answer.
- For Problem 3, is 29 pounds 29 ounces a correct answer? Explain.
- Let's look at how two different students modeled Problem 4. How are they similar? How are they different?
- For Problem 4, how did the drawing of the tape diagram help to find the more efficient way to solve? Why didn't you have to determine Mary's height or the son's height to solve?
- When might it be better to work with the mixed units rather than converting to the smaller unit?
- What are the advantages to knowing several methods for working with units of measurement?

Exit Ticket (3 minutes)

After the Student Debrief, instruct students to complete the Exit Ticket. A review of their work will help with assessing students' understanding of the concepts that were presented in today's lesson and planning more effectively for future lessons. The questions may be read aloud to the students.

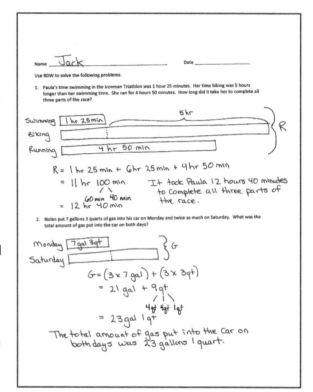

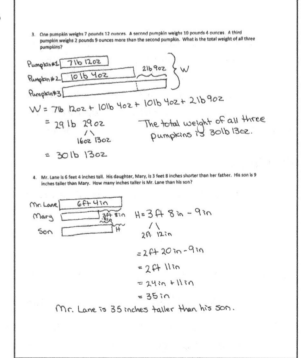

Lesson 10: Solve multi-step measurement word problems.

Name _____ Date _____

Use RDW to solve the following problems.

1. Paula's time swimming in the Ironman Triathlon was 1 hour 25 minutes. Her time biking was 5 hours longer than her swimming time. She ran for 4 hours 50 minutes. How long did it take her to complete all three parts of the race?

2. Nolan put 7 gallons 3 quarts of gas into his car on Monday and twice as much on Saturday. What was the total amount of gas put into the car on both days?

3. One pumpkin weighs 7 pounds 12 ounces. A second pumpkin weighs 10 pounds 4 ounces. A third pumpkin weighs 2 pounds 9 ounces more than the second pumpkin. What is the total weight of all three pumpkins?

4. Mr. Lane is 6 feet 4 inches tall. His daughter, Mary, is 3 feet 8 inches shorter than her father. His son is 9 inches taller than Mary. How many inches taller is Mr. Lane than his son?

Name _____   Date _____

Use RDW to solve the following problem.

Hadley spent 1 hour and 20 minutes completing her math homework, 45 minutes completing her social studies homework, and 30 minutes studying her spelling words. How much time did Hadley spend on homework and studying?

Lesson 10: Solve multi-step measurement word problems.

Name _____ Date _____

Use RDW to solve the following problems.

1. On Saturday, Jeff used 2 quarts 1 cup of water from a full gallon to replace some water that leaked from his fish tank. On Sunday, he used 3 pints of water from the same gallon. How much water was left in the gallon after Sunday?

2. To make punch, Julia poured 1 quart 3 cups of ginger ale into a bowl and then added twice as much fruit juice. How much punch did she make in all?

3. Patti went swimming for 1 hour 15 minutes on Monday. On Tuesday, she swam twice as long as she swam on Monday. On Wednesday, she swam 50 minutes less than the time she swam on Tuesday. How much time did she spend swimming during that three-day period?

4. Myah is 4 feet 2 inches tall. Her sister, Ally, is 10 inches taller. Their little brother is half as tall as Ally. How tall is their little brother in feet and inches?

5. Rick and Laurie have three dogs. Diesel weighs 89 pounds 12 ounces. Ebony weighs 33 pounds 14 ounces less than Diesel. Luna is the smallest at 10 pounds 2 ounces. What is the combined weight of the three dogs in pounds and ounces?

A STORY OF UNITS   Lesson 11 4•7

Lesson 11

Objective: Solve multi-step measurement word problems.

Suggested Lesson Structure

- ■ Fluency Practice   (12 minutes)
- ■ Concept Development   (38 minutes)
- ■ Student Debrief   (10 minutes)
-   **Total Time**   **(60 minutes)**

Fluency Practice (12 minutes)

- Grade 4 Core Fluency Differentiated Practice Sets **4.NBT.4**   (4 minutes)
- Add Mixed Numbers **4.MD.2**   (4 minutes)
- Convert Weight and Time Units **4.MD.1**   (4 minutes)

Grade 4 Core Fluency Differentiated Practice Sets (4 minutes)

Materials: (S) Core Fluency Practice Sets (Lesson 2 Core Fluency Practice Sets)

Note: During Module 7, each day's Fluency Practice may include an opportunity for mastery of the addition and subtraction algorithm by means of the Core Fluency Practice Sets. The process is detailed and Practice Sets are provided in Lesson 2.

Add Mixed Numbers (4 minutes)

Materials: (S) Personal white board

Note: This fluency activity reviews Module 5's fraction work and anticipates today's lesson of adding mixed measurement units. Direct students to respond chorally to the questions or to use written responses on their personal white boards, depending on which is most effective for them.

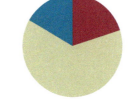

- T: 9 sixteenths + 15 sixteenths is how many sixteenths?
- S: 24 sixteenths.
- T: Express 24 sixteenths as ones and sixteenths.
- S: 1 one and 8 sixteenths.
- T: 9 ounces + 15 ounces is how many ounces?
- S: 24 ounces.
- T: Express 24 ounces as pounds and ounces. Draw a number bond to pull out 16 ounces.
- S: 1 pound 8 ounces.

T: 13 sixteenths + 17 sixteenths is how many sixteenths?
S: 30 sixteenths.
T: Express 30 sixteenths as ones and sixteenths.
S: 1 one and 14 sixteenths.
T: 13 ounces + 17 ounces is how many ounces?
S: 30 ounces.
T: Express 30 ounces as pounds and ounces. Draw a number bond to pull out 16 ounces.

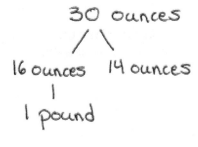

S: 1 pound 14 ounces.

Continue with the following possible sequence: 15 sixteenths + 15 sixteenths related to the same number of ounces.

Convert Weight and Time Units (4 minutes)

Materials: (S) Personal white board

Note: This fluency activity reviews Lessons 1 and 3 and anticipates today's work with weight and time units. Complete as a choral or white board activity.

T: Express each number of pounds and ounces as ounces or days and hours as hours.
T: 1 pound.
S: 16 ounces.
T: 1 pound 10 ounces.
S: 26 ounces.
T: 1 day 2 hours.
S: 26 hours.
T: 2 days 3 hours.
S: 51 hours.

Repeat the same process moving between pounds and ounces and then days and hours.

Concept Development (38 minutes)

Materials: (S) Problem Set

Suggested Delivery of Instruction for Solving Lesson 11's Word Problems

For Problems 1–4, students may work in pairs to solve each of the problems using the RDW approach to problem solving.

1. Model the problem.

Select two pairs of students who can successfully model the problem to work at the board while the other students work independently or in pairs at their seats. Review the following questions before beginning the first problem.

- Can you draw something?
- What can you draw?
- What conclusions can you make from your drawing?

As students work, circulate and reiterate the questions above. After two minutes, have the two pairs of students share only their labeled diagrams. For about one minute, have the demonstrating students receive and respond to feedback and questions from their peers.

2. Calculate to solve and write a statement.

Allow students two minutes to complete work on the problem, sharing their work and thinking with a peer. Have students write their equations and statements of the answer.

3. Assess the solution.

Give students one to two minutes to assess the solutions presented by their peers on the board, comparing the solutions to their own work. Highlight alternative methods to reach the correct solution.

> **NOTES ON MULTIPLE MEANS OF ENGAGEMENT:**
>
> Depending on the needs of English language learners, allow students to discuss their math work in their first language. Alternatively, provide sentence frames and starters such as the ones given below:
>
> - I didn't understand ...
> - Can you explain how ...
> - I thought _____ was a more efficient way of solving because ...

Problem 1

Lauren ran a marathon and finished 1 hour 15 minutes after Amy, who had a time of 2 hours 20 minutes. Cassie finished 35 minutes after Lauren. How long did it take Cassie to run the marathon?

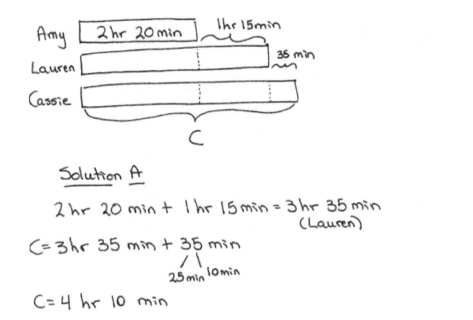

Solution A models solving for Lauren's time and then adding 35 minutes to Lauren's time to solve for Cassie's time. Solution B uses the arrow way to add up, starting with Amy's time of 2 hours 20 minutes and then adding the additional hours and minutes needed to reach Cassie's time. Encourage students to work with the mixed units. However, it should be noted that an answer resulting in 250 minutes is a correct response because it is equivalent to 4 hours 10 minutes. Early finishers can be encouraged to find the sum of their times. The tape diagram shows clearly that we have (3 × 2 hr 20 min) + (2 × 1 hr 15 min + 35 min).

Problem 2

Chef Joe has 8 lb 4 oz of ground beef in his freezer. This is $\frac{1}{3}$ of the amount needed to make the number of burgers he planned for a party. If he uses 4 oz of beef for each burger, how many burgers is he planning to make?

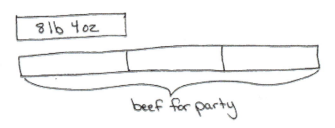

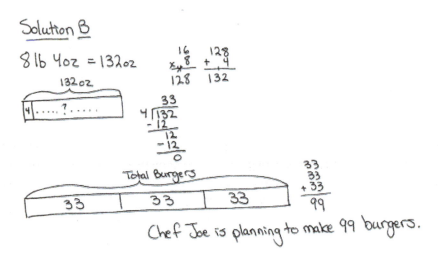

MP.7

Students use their understanding of fractions when they draw a model showing that the total beef needed is made up of three units of 8 pounds 4 ounces (132 ounces) of beef. Addition or multiplication can be used to find the total amount of beef needed. Solution A models solving for the number of burgers made in the total pounds and in the total ounces separately and then adding the number of burgers together. Alternatively, Solution B shows finding the number of burgers that can be made with one-third of the ground beef. Multiplying by three solves for how many burgers can be made with the whole amount of ground beef.

Lesson 11: Solve multi-step measurement word problems.

Problem 3

Sarah read for 1 hour 17 minutes each day for 6 days. If she took 3 minutes to read each page, how many pages did she read in 6 days?

Solution A

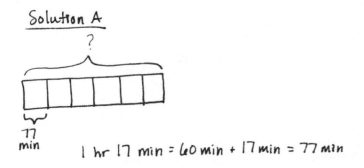

1 hr 17 min = 60 min + 17 min = 77 min

```
  77
×  6
 ---
 462
```
462 min in 6 days

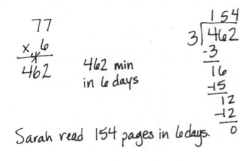

Sarah read 154 pages in 6 days.

Solution B

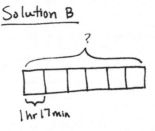

(1 hr × 6) + (17 min × 6) = 6 hr 102 min
                                   = 60 min 42 min
                           = 7 hr 42 min

3 min → 1 page
60 min → 20 pages
7 hr → 140 pages
42 min → 14 pages

140 pages + 14 pages = 154 pages

In 6 days Sarah read 154 pages.

Students may start by converting the mixed units into the minutes read each day. They may then decide to use the minutes read each day to find the pages read each day and then the pages read in six days. Quickly, some may find that solving for the pages read each day results in a remainder, which they may not understand how to interpret. Therefore, encourage students to solve with the whole 6 days in mind. Solution A divides the total number of minutes in 6 days by 3 pages to find that 154 pages are read in 6 days. Solution B finds the mixed units of hours and minutes and solves part to whole, solving for the number of pages in 7 hours and the number of pages in 42 minutes.

Problem 4

Grades 3, 4, and 5 have their annual field day together. Each grade level is given 16 gallons of water. If there are a total of 350 students, will there be enough water for each student to have 2 cups?

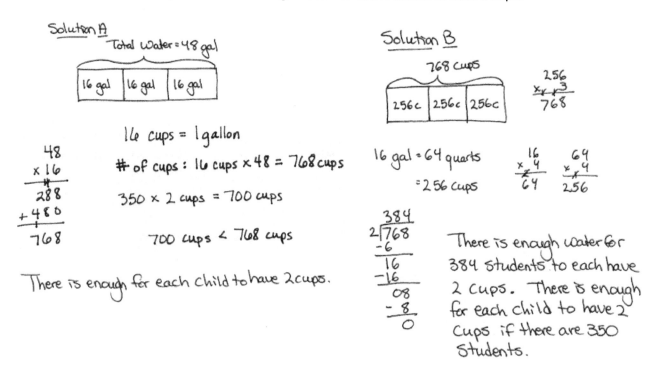

To solve this problem, students must see that each grade level is given 16 gallons of water—or a total of 48 gallons—for field day. Students may choose to convert across two or even three units going from gallons to quarts and quarts to cups as shown in Solution B. Sharing multiple solution strategies can show students the efficiency of using the rules learned in Topic A for converting. Solution A shows the use of the rules by multiplying the number of gallons by 16 to find the number of cups. Comparing the number of cups needed (700 cups) to the number of cups they have for field day (768 cups) allows students to see that they do have enough water for each student to have 2 cups. Alternatively, Solution B divided to find the total number of students who could drink 2 cups of water from 16 gallons of water, proving there is enough water.

Problem Set

Please note that the Problem Set is completed as part of the Concept Development for this lesson.

A STORY OF UNITS   Lesson 11   4•7

Student Debrief (10 minutes)

Lesson Objective: Solve multi-step measurement word problems.

The Student Debrief is intended to invite reflection and active processing of the total lesson experience.

Invite students to review their solutions for the Problem Set. They should check work by comparing answers with a partner before going over answers as a class. Look for misconceptions or misunderstandings that can be addressed in the Debrief. Guide students in a conversation to debrief the Problem Set and process the lesson.

Any combination of the questions below may be used to lead the discussion.

- Why might you want to keep the mixed units in Problem 1? Why might you want to start by converting the mixed units to minutes in Problem 3?
- What challenge might you have faced when solving Problem 3? Why couldn't you first determine the number of pages she read each day?
- If it took Sarah 4 minutes instead of 3 minutes to read a page in Problem 3, would she read more or fewer pages in a week? Explain.
- Some students use strategies that are creative and very different than the majority of the class. How can a student be sure his strategy works?

Exit Ticket (3 minutes)

After the Student Debrief, instruct students to complete the Exit Ticket. A review of their work will help with assessing students' understanding of the concepts that were presented in today's lesson and planning more effectively for future lessons. The questions may be read aloud to the students.

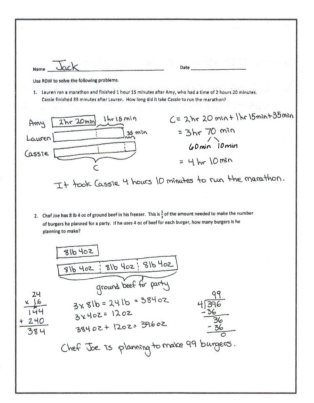

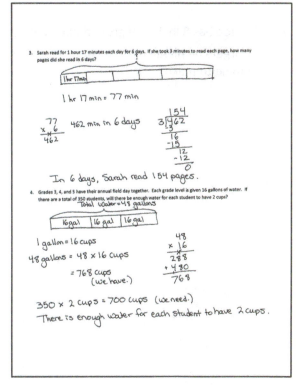

Lesson 11:   Solve multi-step measurement word problems.   147

A STORY OF UNITS

Lesson 11 Problem Set 4•7

Name _____ Date _____

Use RDW to solve the following problems.

1. Lauren ran a marathon and finished 1 hour 15 minutes after Amy, who had a time of 2 hours 20 minutes. Cassie finished 35 minutes after Lauren. How long did it take Cassie to run the marathon?

2. Chef Joe has 8 lb 4 oz of ground beef in his freezer. This is $\frac{1}{3}$ of the amount needed to make the number of burgers he planned for a party. If he uses 4 oz of beef for each burger, how many burgers is he planning to make?

3. Sarah read for 1 hour 17 minutes each day for 6 days. If she took 3 minutes to read each page, how many pages did she read in 6 days?

4. Grades 3, 4, and 5 have their annual field day together. Each grade level is given 16 gallons of water. If there are a total of 350 students, will there be enough water for each student to have 2 cups?

Name _____ Date _____

Use RDW to solve the following problem.

Judy spent 1 hour 15 minutes less than Sandy exercising last week. Sandy spent 50 minutes less than Mary, who spent 3 hours at the gym. How long did Judy spend exercising?

Name _____   Date _____

Use RDW to solve the following problems.

1. Ashley ran a marathon and finished 1 hour 40 minutes after P.J., who had a time of 2 hours 15 minutes. Kerry finished 12 minutes before Ashley. How long did it take Kerry to run the marathon?

2. Mr. Foote's deck is 12 ft 6 in wide. Its length is twice the width plus 3 more inches. How long is the deck?

3. Mrs. Lorentz bought 12 pounds 8 ounces of sugar. This is $\frac{1}{4}$ of the sugar she will use to make sugar cookies in her bakery this week. If she uses 10 ounces of sugar for each batch of sugar cookies, how many batches of sugar cookies will she make in a week?

4. Beth Ann practiced piano for 1 hour 5 minutes each day for 1 week. She had 5 songs to practice and spent the same amount of time practicing each song. How long did she practice each song during the week?

5. The concession stand has 18 gallons of punch. If there are a total of 240 students who want to purchase 1 cup of punch each, will there be enough punch for everyone?

A STORY OF UNITS

Mathematics Curriculum

GRADE 4 • MODULE 7

Topic C
Investigation of Measurements Expressed as Mixed Numbers

4.OA.3, 4.MD.1, 4.MD.2, 4.NBT.5, 4.NBT.6

Focus Standards:	4.OA.3	Solve multi-step word problems posed with whole numbers and having whole-number answers using the four operations, including problems in which remainders must be interpreted. Represent these problems using equations with a letter standing for the unknown quantity. Assess the reasonableness of answers using mental computation and estimation strategies including rounding.
	4.MD.1	Know relative sizes of measurement units within one system of units including km, m, cm; kg, g; lb, oz.; l, ml; hr, min, sec. Within a single system of measurement, express measurements in a larger unit in terms of a smaller unit. Record measurement equivalents in a two-column table. *For example, know that 1 ft is 12 times as long as 1 in. Express the length of a 4 ft snake as 48 in. Generate a conversion table for feet and inches listing the number pairs (1, 12), (2, 24), (3, 36), …*
	4.MD.2	Use the four operations to solve word problems involving distances, intervals of time, liquid volumes, masses of objects, and money, including problems involving simple fractions or decimals, and problems that require expressing measurements given in a larger unit in terms of a smaller unit. Represent measurement quantities using diagrams such as number line diagrams that feature a measurement scale.
Instructional Days:	3	
Coherence -Links from:	G3–M1	Properties of Multiplication and Division and Solving Problems with Units of 2–5 and 10
	G3–M2	Place Value and Problem Solving with Units of Measure
-Links to:	G5–M1	Place Value and Decimal Fractions
	G5–M2	Multi-Digit Whole Number and Decimal Fraction Operations

In Topic C, students convert larger mixed measurement units to smaller units. Students partition a measurement scale in Lesson 12 to help them convert larger units of measurements with fractional parts into smaller units. For example, students use a ruler to draw a number line 1 foot in length. Then, students partition the number line into 12 equal parts. Combining fractions and conversions, students see that 1 twelfth foot is the same as 1 inch. Repeating the same activity—but with different partitions—students find how many inches are in $\frac{1}{2}$ foot, $\frac{1}{3}$ foot, and $\frac{1}{4}$ foot.

Topic C: Investigation of Measurements Expressed as Mixed Numbers

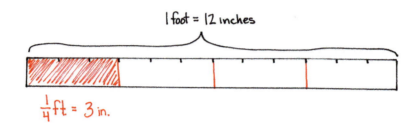

The same hands-on activity can be repeated for the capacity of part of a gallon represented as quarts. That hands-on experience leads students to make abstract connections in Lesson 13 for weight, identifying that $\frac{1}{16}$ pound is equal to 1 ounce, and with respect to time, finding that $\frac{1}{60}$ hour is equal to 1 minute, through the modeling of tape diagrams and number lines. Moving forward, students use their knowledge of conversion tables with this new understanding to convert mixed number units into smaller units, such as $3\frac{1}{4}$ feet equals 39 inches, by applying mixed number units to solve multi-step problems in Lesson 14.

Erin has $1\frac{3}{4}$ pounds of apples. A recipe for apple tarts requires 4 ounces of apples. How many apple tarts can Erin make?

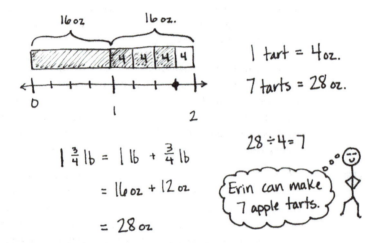

A Teaching Sequence Toward Mastery of Investigation of Measurements Expressed as Mixed Numbers
Objective 1: Use measurement tools to convert mixed number measurements to smaller units. (Lessons 12–13)
Objective 2: Solve multi-step word problems involving converting mixed number measurements to a single unit. (Lesson 14)

A STORY OF UNITS

Lesson 12 4•7

Lesson 12

Objective: Use measurement tools to convert mixed number measurements to smaller units.

Suggested Lesson Structure

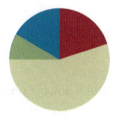

- Fluency Practice   (12 minutes)
- Application Problem   (5 minutes)
- Concept Development   (33 minutes)
- Student Debrief   (10 minutes)
- **Total Time**   **(60 minutes)**

Fluency Practice (12 minutes)

- Grade 4 Core Fluency Differentiated Practice Sets **4.NBT.4**   (4 minutes)
- Complete Length Units **4.MD.1**   (4 minutes)
- Complete One with Fractional Units **4.NF.3a**   (4 minutes)

Grade 4 Core Fluency Differentiated Practice Sets (4 minutes)

Materials: (S) Core Fluency Practice Sets (Lesson 2 Core Fluency Practice Sets)

Note: During Module 7, each day's Fluency Practice may include an opportunity for mastery of the addition and subtraction algorithm by means of the Core Fluency Practice Sets. The process is detailed and Practice Sets are provided in Lesson 2.

Complete Length Units (4 minutes)

Materials: (S) Personal white board

Note: This fluency activity reviews measurement conversions and the important notion of completing the unit.

   T: (Write 2 feet.) How many more feet are needed to make a yard?
   S: (Write 1 foot.)

Continue the complete-the-unit work using the following possible sequence:

- Yards: 1 foot.
- Meters: 50 centimeters, 75 centimeters, 27 centimeters.
- Kilometers: 900 meters, 750 meters, 250 meters, 168 meters.
- Feet: 11 inches, 5 inches, 8 inches.

Lesson 12:   Use measurement tools to convert mixed number measurements to smaller units.

155

A STORY OF UNITS — Lesson 12 4•7

Complete One with Fractional Units (4 minutes)

Materials: (S) Personal white board

Note: This fluency activity reviews fraction work from Module 5 and allows students to see the relationship between measurement and fractional units in anticipation of today's lesson.

T: (Write $\frac{2}{3}$.) How many more thirds complete 1?

S: (Write $\frac{1}{3}$.)

T: (Write $\frac{1}{3}$.) How many more thirds complete 1?

S: (Write $\frac{2}{3}$.)

Continue the complete-the-unit work using the following possible sequence:

$\frac{1}{4}, \frac{2}{4}, \frac{3}{4}, \frac{2}{10}, \frac{5}{10}, \frac{9}{10}, \frac{9}{12}, \frac{10}{12}, \frac{7}{12}, \frac{8}{12}, \frac{8}{16}, \frac{15}{16}, \frac{10}{16}, \frac{10}{100}, \frac{99}{100}, \frac{75}{100}, \frac{75}{1000}, \frac{750}{1000}, \frac{999}{1000}, \frac{1}{1000}$.

Application Problem (5 minutes)

A rectangular tile has a width of 1 foot 6 inches and length of 2 feet. What is the perimeter of the tile?

[Diagram: rectangle, width 1ft 6in, length 2ft]

Solution A
P = 1ft 6in + 1ft 6in + 2ft + 2ft
  = 2ft 12in + 4ft
  = 7ft

Solution B
P = 2(1ft 6in) + 2(2ft)
  = 2ft 12in + 4ft
  = 7ft

The tile has a perimeter of 7 feet.

Note: This Application Problem reviews the mixed unit work from Topic B. In the Debrief, the students can revisit this problem and see that $1\frac{1}{2}$ feet is another way of saying 1 foot 6 inches.

Concept Development (33 minutes)

Materials: (T) 1 gallon container marked to show 4 quarts, 1 gallon container marked to show fourths of a gallon, 1 quart container, colored water (S) 12-inch ruler, yardstick (per group of 3 students), Problem Set, foot-long strip of paper (optional)

Note: Students can work in groups of three so that only 1 yardstick is needed for every three students. Cash register tape can be used for the foot-long strips of paper.

Lesson 12: Use measurement tools to convert mixed number measurements to smaller units.

A STORY OF UNITS   Lesson 12   4•7

Problem 1: Identify $\frac{1}{3}$ yard as 1 foot, and use this equivalence to solve problems.

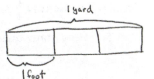

T: 1 yard is how many feet? Use your ruler and yardstick to measure to verify your answer.

S: 3 feet. → 1 yard equals the length of three 1-foot rulers.

T: Look at Problem 1 on your Problem Set. Draw a tape diagram to represent 1 yard decomposed into 3 feet.

S: (Draw tape diagrams.)

T: (Point to 1 unit of the tape diagram.) 1 unit is $\frac{1}{3}$ yard. Why is that?

S: 3 units is 1. $\frac{1}{3} + \frac{1}{3} + \frac{1}{3} = 1$. → 1 is partitioned into 3 units, so 1 unit represents $\frac{1}{3}$.

T: In your group, use your rulers to show $\frac{1}{3}$ yard.

S: (Hold up one ruler against the yardstick.)

T: $\frac{1}{3}$ yard is how many feet?

S: 1 foot.

T: As a group, use your rulers to show $\frac{2}{3}$ yard. $\frac{2}{3}$ yard is how many feet?

S: 2 feet.

T: As a group, use your rulers to show $\frac{3}{3}$ yard. $\frac{3}{3}$ yard is how many feet?

S: 3 feet.

T: Record your responses for Problem 1 on the Problem Set.

T: Talk and work with your partner. How many feet are in $1\frac{2}{3}$ yards? (Allow students time to work.)

S: 5 feet.

T: Explain your thinking.

S: We used our rulers and yardstick and modeled it. → We drew another tape diagram and saw that 2 yards was 6 feet, so 1 third less is 1 foot less, or 5 feet. → We know that 1 yard is 3 feet and $\frac{2}{3}$ yard is 2 feet. 3 and 2 is 5.

T: Draw a tape diagram for Problem 2 on the Problem Set to show that $2\frac{2}{3}$ yards is equal to 8 feet. If you finish early, figure out how many feet are equal to $7\frac{1}{3}$ yards and $35\frac{2}{3}$ yards.

Circulate to check for understanding.

> **NOTES ON MULTIPLE MEANS OF ENGAGEMENT:**
>
> Students working above grade level, and others who convert $1\frac{2}{3}$ yards to feet rapidly and mentally, may model $2\frac{2}{3}$ yards = 8 feet without delay. After doing so, or alternatively, they may like an autonomous partner activity in which they choose their own mixed number amounts of yards that their partners convert and model.

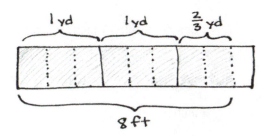

Lesson 12: Use measurement tools to convert mixed number measurements to smaller units.

157

A STORY OF UNITS   Lesson 12 4•7

Problem 2: Identify $\frac{1}{4}$ gallon as 1 quart, and use this equivalence to solve problems.

T: How many quarts equal a gallon?
S: 4 quarts.
T: Yes. This gallon container is marked to show the 4 quarts. (Measure 1 quart from the full gallon and pour it into the gallon container marked to show fourths.) This gallon container is marked to show fourths. One quart of water is in this gallon container. What fraction of a gallon is filled?
S: $\frac{1}{4}$ gallon.
T: $\frac{1}{4}$ gallon is equal to 1 quart. Why?
S: It takes 4 quarts to make 1 gallon. 1 out of 4 parts is $\frac{1}{4}$. → A gallon can be divided many different ways. This one is divided into fourths. We can say the 4 units, or 4 quarts, equals 1 gallon, so $\frac{1}{4}$ gallon is 1 unit or 1 quart.

Repeat by pouring additional units of 1 quart of water into the gallon container, asking for the fraction of a gallon being represented after each addition. Be sure to elicit the equivalence of $\frac{2}{4}$ gallon and $\frac{1}{2}$ gallon.

T: Draw a tape diagram to show 4 quarts equals 1 gallon.
S: (Draw as shown to the right.)
T: We have divided the gallon into 4 equal parts. What fraction represents 1 quart?
S: $\frac{1}{4}$.
T: Draw the tape diagrams for Problems 3 and 4 of your Problem Set.

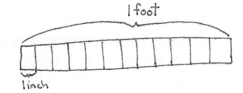

Circulate to check for student understanding.

Problem 3: Identify $\frac{1}{12}$ foot as 1 inch, and use this equivalence to solve problems.

T: Look at your rulers. 1 foot equals how many inches?
S: 12 inches.
T: Draw a tape diagram where the tape represents 1 foot and each unit represents 1 inch.
S: (Draw a tape diagram with 12 units, as shown to the right.)

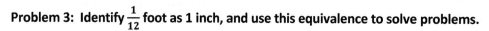

Note: If time management is not a concern, invite students to mark inches on a foot-long strip and to fold the paper before drawing the tape diagram.

T: 1 unit represents 1 inch. $\frac{1}{12}$ foot equals how many inches? Tell me the complete number sentence.
S: $\frac{1}{12}$ foot = 1 inch.
T: $\frac{2}{12}$ foot?
S: 2 inches.

158   Lesson 12:   Use measurement tools to convert mixed number measurements to smaller units.

Have partners quickly proceed with the pattern up to 12 twelfths for 1 foot.

- T: Some of these fractions can be expressed in larger units. Shade 1 half foot of your tape diagram.
- S: (Shade the tape diagram.)
- T: How many inches are equal to $\frac{1}{2}$ foot?
- S: 6 inches!
- T: (Write $\frac{1}{2}$ ft = $\frac{}{12}$ ft.) Talk to your partner. Instead of just using the tape diagram, how can we use what we know about finding equivalent fractions to find the number of twelfths equal to $\frac{1}{2}$ foot?
- S: I know 2 times 6 is 12, so I can multiply the numerator by the same factor: $\frac{1 \times 6}{2 \times 6} = \frac{6}{12}$. → It's like a number line. A half is decomposed into 6 smaller parts: $\frac{1}{2}$ foot = $\frac{6}{12}$ foot. → We are making larger units. Six inches is now a unit, so we have 1 out of 2 units, one-half.
- T: Again, how many inches are equal to $\frac{1}{2}$ or $\frac{6}{12}$ foot?
- S: 6 inches.
- T: Work with your partner to find how many inches are equal to $\frac{1}{4}$ foot. (Allow students time to work.)
- T: How did you figure it out?
- S: To find one fourth, we just cut the half in half on the tape diagram to see 1 fourth is equal to 3 inches. → We also set it up as an equivalent fraction, $\frac{1}{4} = \frac{}{12}$. Four times 3 is 12, so that meant the numerator would be 3, too.

If students need more guidance, repeat the same process with $\frac{1}{3}$ foot.

- T: Talk to your partner. How many inches are equal to $4\frac{1}{2}$ feet?
- S: Easy. 4 × 12 is 48. I know $\frac{1}{2}$ foot is 6 inches, so 48 + 6 = 54. There are 54 inches. → We drew a tape diagram with 5 equal units but partitioned the last unit in half.

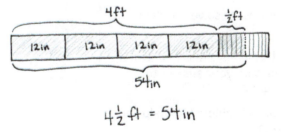

Repeat, using multiplication to find the equivalent number of twelfths in $\frac{3}{4}$ foot, the number of inches in $\frac{3}{4}$ foot, and the number of inches in $2\frac{3}{4}$ feet.

- T: Solve Problem 5(a–f) and Problem 6 on your Problem Set using equivalent fractions or a tape diagram.

Circulate to check for student understanding.

A STORY OF UNITS

Lesson 12

Problem Set (10 minutes)

Students should do their personal best to complete the remainder of the Problem Set within the allotted 10 minutes. For some classes, it may be appropriate to modify the assignment by specifying which problems they work on first. Some problems do not specify a method for solving. Students should solve these problems using the RDW approach used for Application Problems.

Student Debrief (10 minutes)

Lesson Objective: Use measurement tools to convert mixed number measurements to smaller units.

The Student Debrief is intended to invite reflection and active processing of the total lesson experience.

Invite students to review their solutions for the Problem Set. They should check work by comparing answers with a partner before going over answers as a class. Look for misconceptions or misunderstandings that can be addressed in the Debrief. Guide students in a conversation to debrief the Problem Set and process the lesson.

Any combination of the questions below may be used to lead the discussion.

- How is Problem 1(a), $\frac{1}{3}$ yard = 1 foot, a similar statement to Problem 5(a), $\frac{1}{12}$ foot = 1 inch?
- Explain to your partner how to solve Problem 6(b).
- How can knowing that 8 gallons equals 32 quarts help you check to make sure your answer to Problem 6(d) is reasonable?
- How could your answer to Problem 6(g) help you figure out Problem 6(h)?
- How could we rewrite the dimensions of the tile from the Application Problem using a mixed number instead of mixed units of feet and inches?

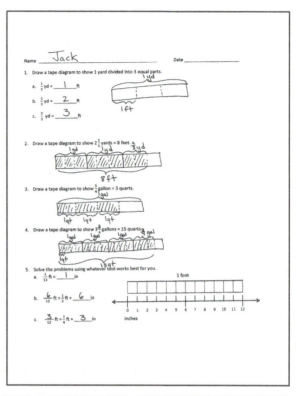

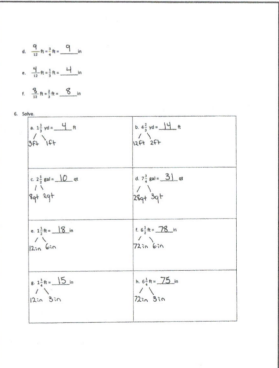

Lesson 12: Use measurement tools to convert mixed number measurements to smaller units.

Exit Ticket (3 minutes)

After the Student Debrief, instruct students to complete the Exit Ticket. A review of their work will help with assessing students' understanding of the concepts that were presented in today's lesson and planning more effectively for future lessons. The questions may be read aloud to the students.

Name _____   Date _____

1. Draw a tape diagram to show 1 yard divided into 3 equal parts.

    a. $\frac{1}{3}$ yd = _____ ft

    b. $\frac{2}{3}$ yd = _____ ft

    c. $\frac{3}{3}$ yd = _____ ft

2. Draw a tape diagram to show $2\frac{2}{3}$ yards = 8 feet.

3. Draw a tape diagram to show $\frac{3}{4}$ gallon = 3 quarts.

4. Draw a tape diagram to show $3\frac{3}{4}$ gallons = 15 quarts.

5. Solve the problems using whatever tool works best for you.

    a. $\frac{1}{12}$ ft = _____ in

    b. 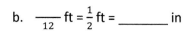 ft = $\frac{1}{2}$ ft = _____ in

    c. 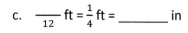 ft = $\frac{1}{4}$ ft = _____ in

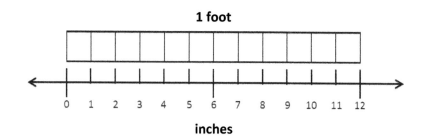

1 foot

inches

Lesson 12: Use measurement tools to convert mixed number measurements to smaller units.

d. $\frac{}{12}$ ft = $\frac{3}{4}$ ft = _____ in

e. $\frac{}{12}$ ft = $\frac{1}{3}$ ft = _____ in

f. $\frac{}{12}$ ft = $\frac{2}{3}$ ft = _____ in

6. Solve.

a. $1\frac{1}{3}$ yd = _____ ft	b. $4\frac{2}{3}$ yd = _____ ft
c. $2\frac{1}{2}$ gal = _____ qt	d. $7\frac{3}{4}$ gal = _____ qt
e. $1\frac{1}{2}$ ft = _____ in	f. $6\frac{1}{2}$ ft = _____ in
g. $1\frac{1}{4}$ ft = _____ in	h. $6\frac{1}{4}$ ft = _____ in

Name _____   Date _____

1. Solve the problems using whatever tool works best for you.

    a. $\dfrac{}{12}$ ft = $\dfrac{1}{2}$ ft = _____ in

    b. $\dfrac{}{12}$ ft = $\dfrac{3}{4}$ ft = _____ in

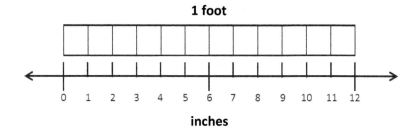

2. Solve.

    a. $1\dfrac{1}{3}$ yd = _____ ft

    b. $5\dfrac{3}{4}$ gal = _____ qt

A STORY OF UNITS   Lesson 12 Homework   4•7

Name _____   Date _____

1. Draw a tape diagram to show $1\frac{1}{3}$ yards = 4 feet.

2. Draw a tape diagram to show $\frac{1}{2}$ gallon = 2 quarts.

3. Draw a tape diagram to show $1\frac{3}{4}$ gallons = 7 quarts.

4. Solve the problems using whatever tool works best for you.

   a. $\frac{1}{2}$ foot = _____ inches

   b. $\frac{}{12}$ foot = $\frac{1}{4}$ foot = _____ inches

   c. $\frac{}{12}$ foot = $\frac{1}{6}$ foot = _____ inches

   d. $\frac{}{12}$ foot = $\frac{1}{3}$ foot = _____ inches

   e. $\frac{}{12}$ foot = $\frac{2}{3}$ foot = _____ inches

   f. $\frac{}{12}$ foot = $\frac{5}{6}$ foot = _____ inches

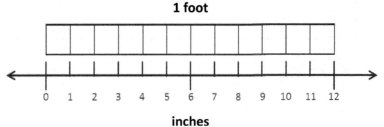

Lesson 12: Use measurement tools to convert mixed number measurements to smaller units.

5. Solve.

a. $2\frac{2}{3}$ yd = _____ ft	b. $3\frac{1}{3}$ yd = _____ ft
c. $3\frac{1}{2}$ gal = _____ qt	d. $5\frac{1}{4}$ gal = _____ qt
e. $6\frac{1}{4}$ ft = _____ in	f. $7\frac{1}{3}$ ft = _____ in
g. $2\frac{1}{2}$ ft = _____ in	h. $5\frac{3}{4}$ ft = _____ in
i. $9\frac{2}{3}$ ft = _____ in	j. $7\frac{5}{6}$ ft = _____ in

Lesson 12: Use measurement tools to convert mixed number measurements to smaller units.

Lesson 13

Objective: Use measurement tools to convert mixed number measurements to smaller units.

Suggested Lesson Structure

■ Fluency Practice   (10 minutes)
■ Application Problem   (5 minutes)
■ Concept Development   (35 minutes)
■ Student Debrief   (10 minutes)
  Total Time   **(60 minutes)**

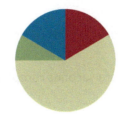

Fluency Practice (10 minutes)

- Grade 4 Core Fluency Differentiated Practice Sets **4.NBT.4**   (4 minutes)
- Complete Time Units **4.MD.1**   (3 minutes)
- Complete Weight Units **4.MD.1**   (3 minutes)

Grade 4 Core Fluency Differentiated Practice Sets (4 minutes)

Materials: (S) Core Fluency Practice Sets (Lesson 2 Core Fluency Practice Sets)

Note: During Module 7, each day's Fluency Practice may include an opportunity for mastery of the addition and subtraction algorithm by means of the Core Fluency Practice Sets. The process is detailed and Practice Sets are provided in Lesson 2.

Complete Time Units (3 minutes)

Materials: (S) Personal white board

Note: This fluency activity reviews Lesson 3. Depending on the class, students might write responses on their personal white boards or respond orally.

  T: (Write 4 days.) How many more days complete the week?
  S: 3 days.
  T: (Write 40 min.) How many more minutes complete the hour?
  S: 20 minutes.

> **NOTES ON MULTIPLE MEANS OF ACTION AND EXPRESSION:**
>
> During the Complete Time Units fluency activity, couple writing and speaking. For example, say "four days" while writing 4 days on the personal white board. This helps both English language learners and others as they process the activity.
>
> Also, to support students working below grade level, present equivalencies, such as 24 hours = 1 day, that students may refer to as they solve.

A STORY OF UNITS   Lesson 13   4•7

T: (Write 25 min.) How many more minutes complete the hour?
S: 35 minutes.
T: (Write 18 min.) How many more minutes complete the hour?
S: 42 minutes.
T: (Write 18 hours.) How many more hours complete the day?
S: 6 hours.
T: (Write 10 hours.) How many more hours complete the day?
S: 14 hours.
T: (Write 20 seconds.) How many more seconds complete the minute?
S: 40 seconds.
T: (Write 34 seconds.) How many more seconds complete the minute?
S: 26 seconds.

Complete Weight Units (3 minutes)

Materials: (S) Personal white board

Note: This fluency activity reviews measurement conversions from Lesson 1 and the important concept of completing the unit.

T: (Write 15 ounces.) How many more ounces complete the pound?
S: (Write 1 ounce.)

Continue the complete-the-unit work using the following possible sequence: 8 ounces, 12 ounces, 4 ounces, and 7 ounces.

Application Problem (5 minutes)

Micah used $3\frac{3}{4}$ gallons of paint to paint his bathroom. He used 3 times as much paint to paint his bedroom. How many quarts of paint did it take to paint his bedroom?

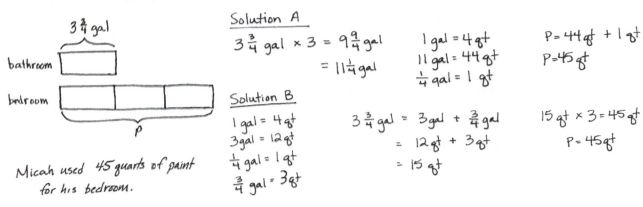

Lesson 13: Use measurement tools to convert mixed number measurements to smaller units.

A STORY OF UNITS   Lesson 13   4•7

Note: Reviewing Lesson 12, students use multiplicative reasoning and the conversion of mixed number measurements to solve this problem. Solution A solves for the total number of gallons and then converts to quarts. Solution B finds how many quarts equal 1 unit and then multiplies to find how many quarts are in 3 units.

Concept Development (35 minutes)

Materials: (S) Problem Set

Problem 1: Identify $\frac{1}{16}$ pound as 1 ounce.

T: 1 pound is equal to how many ounces?

S: 16 ounces.

T: Draw a tape diagram to represent 1 pound. You said that 16 ounces equals 1 pound. Show this on your tape diagram.

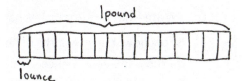

S: (Draw the tape diagram for 1 pound.)

T: $\frac{1}{16}$ pound equals how many ounces? Tell me the complete number sentence.

S: $\frac{1}{16}$ pound = 1 ounce.

T: $\frac{2}{16}$ pound equals how many ounces?

S: 2 ounces.

T: Find the number of ounces equal to $\frac{1}{2}$ pound. Explain your thinking to your partner.

S: $\frac{1}{2}$ is equal to $\frac{8}{16}$, so $\frac{1}{2}$ pound must be 8 ounces. → I shaded half the tape diagram and saw it was equal to 8 units. A unit is an ounce, so $\frac{1}{2}$ pound = 8 ounces. → We used equivalent fractions: $\frac{1 \times 8}{2 \times 8} = \frac{8}{16}$.

T: How does the number line next to Problem 1 on your Problem Set illustrate this fact?

S: After shading, it's easy to see the halfway point on the tape diagram lines up with 8 ounces on the number line. → You can see that a half is 8 ounces, or 8 of the small length units. It might be even clearer if we labeled 0 and 16 as 0 pounds and 1 pound not just 0 ounces and 16 ounces.

T: With a partner, complete Problems 1 and 2 on your Problem Set.

Ask early finishers to find the number of ounces in $8\frac{3}{4}$ pounds and $11\frac{1}{2}$ pounds. Circulate and provide support as necessary.

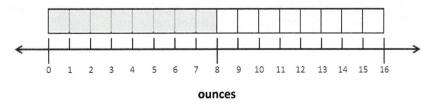

ounces

Lesson 13: Use measurement tools to convert mixed number measurements to smaller units.   169

Problem 2: Identify $\frac{1}{60}$ hour as 1 minute.

T: 1 hour equals 60 minutes. $\frac{1}{60}$ hour equals how many minutes?

S: 1 minute.

T: Discuss with your partner. How many minutes are in $\frac{1}{2}$ hour?

S: 30 minutes. → I know that when the minute hand has gone halfway around the clock, it's 30 minutes.

T: (Write $\frac{1}{2}$ hour = $\frac{}{60}$ hour.) What equivalent fraction could we write to show how many sixtieths of an hour equal $\frac{1}{2}$ hour? Use multiplication to show the equivalence.

S: $\frac{1}{2} = \frac{1 \times 30}{2 \times 30} = \frac{30}{60}$.

T: (Write $\frac{1}{4}$ hour = $\frac{}{60}$ hour.) Determine with your partner how to find the number of minutes in a quarter of an hour.

S: I could take the number of minutes in a half an hour and divide them in half to get 15 minutes. → I know that when the minute hand has gone a quarter of the way around the clock, 15 minutes have passed. → $\frac{1}{4} = \frac{1 \times 15}{4 \times 15} = \frac{15}{60}$.

MP.3

T: How many minutes are there in $3\frac{1}{2}$ hours? Use whatever strategy helps you.

S: 210 minutes.

If time allows, ask students to find the number of minutes in $8\frac{1}{4}$ hours and the number of hours in $8\frac{1}{4}$ days. Have students start the remainder of the Problem Set as soon as they show they can apply their learning to solve the number of smaller units in a mixed number.

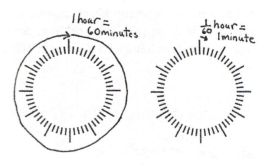

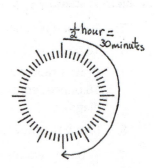

NOTES ON MULTIPLE MEANS OF ENGAGEMENT:

Students working above grade level, and others who convert $3\frac{1}{2}$ hours to minutes rapidly and mentally, may enjoy an autonomous partner activity in which they choose their own mixed number amounts of hours that their partners will convert to minutes, or more challenging, seconds.

Problem Set (10 minutes)

Students should do their personal best to complete the remainder of the Problem Set within the allotted 10 minutes. For some classes, it may be appropriate to modify the assignment by specifying which problems they work on first. Some problems do not specify a method for solving. Students should solve these problems using the RDW approach used for Application Problems.

A STORY OF UNITS   Lesson 13   4•7

Student Debrief (10 minutes)

Lesson Objective: Use measurement tools to convert mixed number measurements to smaller units.

The Student Debrief is intended to invite reflection and active processing of the total lesson experience.

Invite students to review their solutions for the Problem Set. They should check work by comparing answers with a partner before going over answers as a class. Look for misconceptions or misunderstandings that can be addressed in the Debrief. Guide students in a conversation to debrief the Problem Set and process the lesson.

Any combination of the questions below may be used to lead the discussion.

- How could your answer to Problem 5(a) help you solve 5(b)?
- Explain to your partner how to solve Problem 5(i). How do you know that your answer is reasonable?
- How does knowing that 5 × 12 equals 60 and 6 × 12 equals 72 help you see that your answer to Problem 5(m) is reasonable?
- What is the advantage of saying $3\frac{9}{12}$ feet rather than $3\frac{3}{4}$ feet?
- When have you heard someone talk about a fraction of a unit before? Think of examples using the units we have worked with today along with other units of measurement.

Exit Ticket (3 minutes)

After the Student Debrief, instruct students to complete the Exit Ticket. A review of their work will help with assessing students' understanding of the concepts that were presented in today's lesson and planning more effectively for future lessons. The questions may be read aloud to the students.

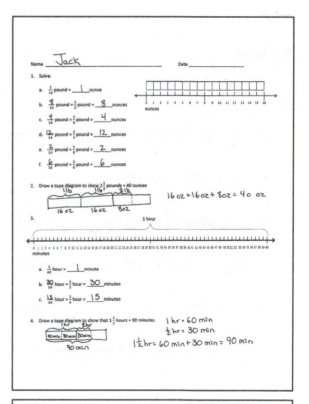

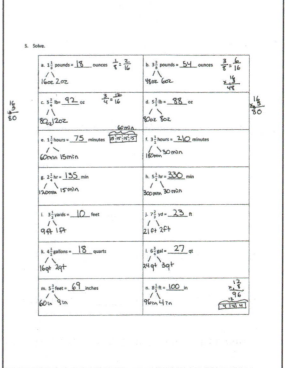

Lesson 13: Use measurement tools to convert mixed number measurements to smaller units.

171

Name _____ Date _____

1. Solve.

   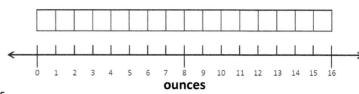

   a. $\frac{1}{16}$ pound = _____ ounce

   b. $\frac{\square}{16}$ pound = $\frac{1}{2}$ pound = _____ ounces

   c. $\frac{\square}{16}$ pound = $\frac{1}{4}$ pound = _____ ounces

   d. $\frac{\square}{16}$ pound = $\frac{3}{4}$ pound = _____ ounces

   e. $\frac{\square}{16}$ pound = $\frac{1}{8}$ pound = _____ ounces

   f. $\frac{\square}{16}$ pound = $\frac{3}{8}$ pound = _____ ounces

2. Draw a tape diagram to show $2\frac{1}{2}$ pounds = 40 ounces.

3.

   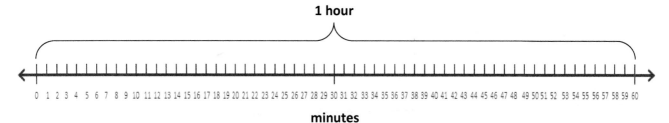

   a. $\frac{1}{60}$ hour = _____ minute

   b. $\frac{\square}{60}$ hour = $\frac{1}{2}$ hour = _____ minutes

   c. $\frac{\square}{60}$ hour = $\frac{1}{4}$ hour = _____ minutes

4. Draw a tape diagram to show that $1\frac{1}{2}$ hours = 90 minutes.

Lesson 13 Problem Set 4•7

5. Solve.

a. $1\frac{1}{8}$ pounds = _____ ounces	b. $3\frac{3}{8}$ pounds = _____ ounces
c. $5\frac{3}{4}$ lb = _____ oz	d. $5\frac{1}{2}$ lb = _____ oz
e. $1\frac{1}{4}$ hours = _____ minutes	f. $3\frac{1}{2}$ hours = _____ minutes
g. $2\frac{1}{4}$ hr = _____ min	h. $5\frac{1}{2}$ hr = _____ min
i. $3\frac{1}{3}$ yards = _____ feet	j. $7\frac{2}{3}$ yd = _____ ft
k. $4\frac{1}{2}$ gallons = _____ quarts	l. $6\frac{3}{4}$ gal = _____ qt
m. $5\frac{3}{4}$ feet = _____ inches	n. $8\frac{1}{3}$ ft = _____ in

Lesson 13: Use measurement tools to convert mixed number measurements to smaller units.

A STORY OF UNITS   Lesson 13 Exit Ticket   4•7

Name _____   Date _____

1. Draw a tape diagram to show that $4\frac{3}{4}$ gallons = 19 quarts.

2. Solve.

a. $1\frac{1}{4}$ pounds = _____ ounces	b. $2\frac{3}{4}$ hr = _____ min
c. $5\frac{1}{2}$ feet = _____ inches	d. $3\frac{5}{6}$ ft = _____ in

Name _____ Date _____

1. Solve.

   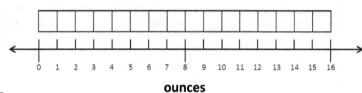
   ounces

   a. $\frac{1}{16}$ pound = _____ ounce

   b. $\frac{}{16}$ pound = $\frac{1}{2}$ pound = _____ ounces

   c. $\frac{}{16}$ pound = $\frac{1}{4}$ pound = _____ ounces

   d. $\frac{}{16}$ pound = $\frac{3}{4}$ pound = _____ ounces

   e. $\frac{}{16}$ pound = $\frac{1}{8}$ pound = _____ ounces

   f. $\frac{}{16}$ pound = $\frac{5}{8}$ pound = _____ ounces

2. Draw a tape diagram to show $1\frac{1}{4}$ pounds = 20 ounces.

3. Solve.

   1 hour

   minutes

   a. $\frac{1}{60}$ hour = _____ minute

   b. $\frac{}{60}$ hour = $\frac{1}{2}$ hour = _____ minutes

   c. $\frac{}{60}$ hour = $\frac{1}{4}$ hour = _____ minutes

   d. $\frac{}{60}$ hour = $\frac{1}{3}$ hour = _____ minutes

4. Draw a tape diagram to show that $2\frac{1}{4}$ hours = 135 minutes.

Lesson 13: Use measurement tools to convert mixed number measurements to smaller units.

5. Solve.

a. $2\frac{1}{4}$ pounds = _____ ounces	b. $4\frac{7}{8}$ pounds = _____ ounces
c. $6\frac{3}{4}$ lb = _____ oz	d. $4\frac{1}{8}$ lb = _____ oz
e. $1\frac{3}{4}$ hours = _____ minutes	f. $4\frac{1}{2}$ hours = _____ minutes
g. $3\frac{3}{4}$ hr = _____ min	h. $5\frac{1}{3}$ hr = _____ min
i. $4\frac{2}{3}$ yards = _____ feet	j. $6\frac{1}{3}$ yd = _____ ft
k. $4\frac{1}{4}$ gallons = _____ quarts	l. $2\frac{3}{4}$ gal = _____ qt
m. $6\frac{1}{4}$ feet = _____ inches	n. $9\frac{5}{6}$ ft = _____ in

Lesson 13: Use measurement tools to convert mixed number measurements to smaller units.

Lesson 14

Objective: Solve multi-step word problems involving converting mixed number measurements to a single unit.

Suggested Lesson Structure

- ■ Fluency Practice   (9 minutes)
- ■ Concept Development   (41 minutes)
- ■ Student Debrief   (10 minutes)
- **Total Time**   **(60 minutes)**

Fluency Practice (9 minutes)

- Complete Length Units **4.MD.1**   (3 minutes)
- Complete Weight Units **4.MD.1**   (3 minutes)
- Complete Capacity Units **4.MD.1**   (3 minutes)

Complete Length Units (3 minutes)

Materials: (S) Personal white board

Note: This fluency activity reviews measurement conversions and the important concept of completing the unit.

- T: (Write 90 centimeters.) How many more centimeters complete 1 meter?
- S: (Write 10 centimeters.)

Continue the complete-the-unit work using the following possible sequence:

- Meters: 50 centimeters, 25 centimeters, 36 centimeters
- Yards: 1 foot
- Kilometers: 500 meters, 650 meters, 350 meters, 479 meters
- Feet: 10 inches, 6 inches, 7 inches

Complete Weight Units (3 minutes)

Materials: (S) Personal white board

Note: This fluency activity reviews measurement conversions and the important concept of completing the unit.

- T: (Write 10 ounces.) How many more ounces complete 1 pound?
- S: (Write 6 ounces.)

Lesson 14: Solve multi-step word problems involving converting mixed number measurements to a single unit.

Continue the complete-the-unit work using the following possible sequence:

- Pounds: 8 ounces
- Kilograms: 900 grams, 750 grams, 250 grams, 378 grams

Complete Capacity Units (3 minutes)

Materials: (S) Personal white board

Note: This fluency activity reviews measurement conversions and the important concept of completing the unit.

T: (Write 3 quarts.) How many more quarts complete 1 gallon?
S: (Write 1 quart.)

Continue the complete-the-unit work using the following possible sequence:

- Gallons: 1 quart
- Liters: 500 milliliters, 200 milliliters, 850 milliliters, 647 milliliters
- Quarts: 2 cups, 1 cup

Concept Development (41 minutes)

Materials: (S) Problem Set

Suggested Delivery of Instruction for Solving Lesson 14's Word Problems

For Problems 1–4, students may work in pairs to solve each of the problems using the RDW approach to problem solving.

1. Model the problem.

Select two pairs of students who can successfully model the problem to work at the board while the other students work independently or in pairs at their seats. Review the following questions before beginning the first problem.

- Can you draw something?
- What can you draw?
- What conclusions can you make from your drawing?

As students work, circulate and reiterate the questions above. After two minutes, have the two pairs of students share only their labeled diagrams. For about one minute, have the demonstrating students receive and respond to feedback and questions from their peers.

2. Calculate to solve and write a statement.

Allow students two minutes to complete work on the problem, sharing their work and thinking with a peer. Have students write their equations and statements of the answer.

3. **Assess the solution.**

Give students one to two minutes to assess the solutions presented by their peers on the board, comparing the solutions to their own work. Highlight alternative methods to reach the correct solution.

Problem 1

A cartoon lasts $\frac{1}{2}$ hour. A movie is 6 times as long as the cartoon. How many minutes does it take to watch both the cartoon and the movie?

> **NOTES ON MULTIPLE MEANS OF ENGAGEMENT:**
>
> Instead of solving Problem 1, allow students working above grade level to solve the remainder of the problems, and then offer an open-ended challenge. Ask students to complete one of the following alternatives:
>
> - Use the model from one of the problems to write a word problem of your own. Be sure to include measurement unit conversions.
> - Write a script to explain your strategy for solving one problem. Share your script with your partner during the Debrief.

Solution A

$\frac{1}{2} = \frac{30}{60}$

30 minutes × 7 = 210 minutes

Solution C

1 unit = $\frac{1}{2}$ hour

7 units = 7 × $\frac{1}{2}$ hr = $\frac{7}{2}$ hr = $3\frac{1}{2}$ hr

$\frac{1}{2}$ hr = 30 min
3 hr = 180 min
$3\frac{1}{2}$ hr = 180 min + 30 min = 210 min

Solution B

$\frac{1}{2} \times 7 = \frac{7}{2} = 3\frac{1}{2}$

$\frac{7}{2} = \frac{6}{2} + \frac{1}{2}$

$3\frac{1}{2}$ hours

$(3 \times 60) + 30 = 210$

It takes 210 minutes to watch both the cartoon and movie.

In Solution A, the student determines that one unit is 30 minutes using equivalent fractions and multiplies by 7 to find the total number of minutes. In both Solutions B and C, the student multiplies $\frac{1}{2}$ by 7 to get $3\frac{1}{2}$ hours and then determines the total number of minutes. Solution B shows numerical work without noting the measurement units, re-contextualizing the 210 by writing the statement. Solution C keeps the measurement units throughout the process, figuring out the number of minutes in a half hour and adding them to the number of minutes in 3 hours.

Problem 2

A large bench is $7\frac{1}{6}$ feet long. It is 17 inches longer than a shorter bench. How many inches long is the shorter bench?

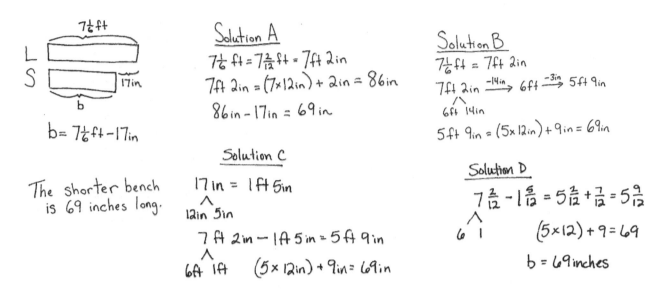

Equivalent fractions are used to determine that $\frac{1}{6}$ foot equals 2 inches in Solution A. The length of the longer bench is converted from feet into inches. To find the length of the shorter bench, 17 inches is subtracted from the length of the longer bench. In Solution B, the student decomposes 7 feet 2 inches into 6 feet 14 inches and subtracts. First, 14 inches are subtracted to get to 6 feet. Next, the student subtracts the remaining 3 inches to get 5 feet 9 inches. This mixed unit measurement is then converted into 69 inches to find the length of the shorter bench in inches. Solution C shows converting the difference in length into a mixed unit to subtract. Solution D shows converting both to mixed numbers to subtract. Each is then converted to inches.

Problem 3

The first container holds 4 gallons 2 quarts of juice. The second container can hold $1\frac{3}{4}$ gallons more than the first container. Altogether, how much juice can the two containers hold?

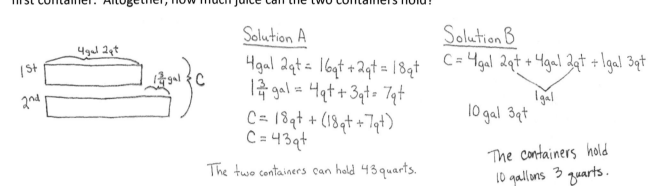

Solution A is solved by converting the mixed unit and the mixed number measurement into quarts, converting to a smaller unit. Remembering 4 quarts equals 1 gallon, students solve for $1\frac{3}{4}$ gallons as 7 quarts. The total number of quarts in the two containers can then be found by adding the number of quarts held by the first and second containers together. In Solution B, the units are kept as mixed measurements. The tape diagram makes it easy to see the total is made up of two equal parts of 4 gallons 2 quarts plus 1 gallon 3 quarts because the second container is $1\frac{3}{4}$ gallons more than the first. This results in a total of 10 gallons 3 quarts that is then converted into 43 quarts.

Problem 4

A girl's height is $3\frac{1}{3}$ feet. A giraffe's height is 3 times that of the girl's. How many inches taller is the giraffe than the girl?

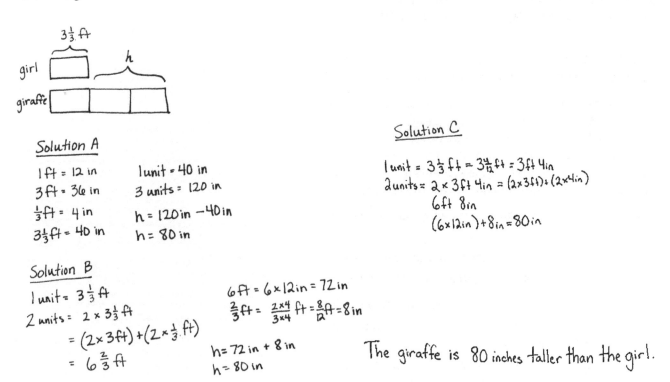

Solution A converts the height of the girl into inches. Then, the height of the giraffe is found by solving for 3 units, as 1 unit equals 40 inches. The heights are subtracted to find the difference of 80 inches. Once the model is drawn, students may see that they do not need to find the giraffe's height to solve this problem. Both Solutions B and C recognize that the two units in the model represent how much taller the giraffe is than the girl. Solution C uses the distributive property to multiply the mixed number by 2. Solution C converts the feet into inches.

A STORY OF UNITS　　　　　　　　　　　　　　　　　　　　　　　　　Lesson 14   4•7

Problem Set

Please note that Problems 1–4 on the Problem Set are completed during instruction. As students present themselves ready to solve independently, release some students or the entire class to work independently or in partnerships. Problems 5–6 can be completed individually, with a partner, or via whole-class instruction.

Student Debrief (10 minutes)

Lesson Objective: Solve multi-step word problems involving converting mixed number measurements to a single unit.

The Student Debrief is intended to invite reflection and active processing of the total lesson experience.

Invite students to review their solutions for the Problem Set. They should check work by comparing answers with a partner before going over answers as a class. Look for misconceptions or misunderstandings that can be addressed in the Debrief. Guide students in a conversation to debrief the Problem Set and process the lesson.

Any combination of the questions below may be used to lead the discussion.

- In Problem 1, how many different ways were 7 halves represented? (30 min × 7, as $\frac{7}{2}$ and as $\frac{6}{2}+\frac{1}{2}$.) What advantage is there to knowing all of these representations when it comes to solving a problem like this one?
- Explain to your partner how you solved Problem 2. If you used different strategies, discuss how you arrived at the same answer.
- What shortcuts or efficiencies did you use today when solving your problems? How do you decide whether to start by converting to a smaller unit or to work with the mixed number measurements?
- How is the remainder in Problem 5 interpreted?
- Did you have trouble persevering at times? When? What can you do to stay focused?

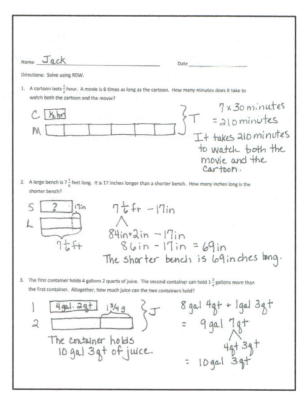

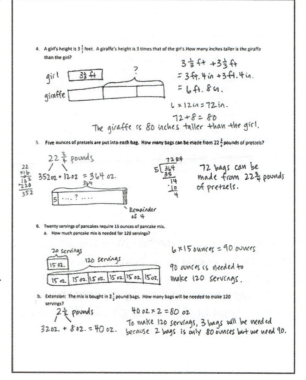

182　　Lesson 14:　Solve multi-step word problems involving converting mixed number measurements to a single unit.

Exit Ticket (3 minutes)

After the Student Debrief, instruct students to complete the Exit Ticket. A review of their work will help with assessing students' understanding of the concepts that were presented in today's lesson and planning more effectively for future lessons. The questions may be read aloud to the students.

A STORY OF UNITS   Lesson 14 Problem Set 4•7

Name _____   Date _____

Use RDW to solve the following problems.

1. A cartoon lasts $\frac{1}{2}$ hour. A movie is 6 times as long as the cartoon. How many minutes does it take to watch both the cartoon and the movie?

2. A large bench is $7\frac{1}{6}$ feet long. It is 17 inches longer than a shorter bench. How many inches long is the shorter bench?

3. The first container holds 4 gallons 2 quarts of juice. The second container can hold $1\frac{3}{4}$ gallons more than the first container. Altogether, how much juice can the two containers hold?

Lesson 14: Solve multi-step word problems involving converting mixed number measurements to a single unit.

4. A girl's height is $3\frac{1}{3}$ feet. A giraffe's height is 3 times that of the girl's. How many inches taller is the giraffe than the girl?

5. Five ounces of pretzels are put into each bag. How many bags can be made from $22\frac{3}{4}$ pounds of pretzels?

6. Twenty servings of pancakes require 15 ounces of pancake mix.

   a. How much pancake mix is needed for 120 servings?

   b. Extension: The mix is bought in $2\frac{1}{2}$-pound bags. How many bags will be needed to make 120 servings?

A STORY OF UNITS

Lesson 14 Exit Ticket 4•7

Name _____ Date _____

Use RDW to solve the following problem.

It took Gigi 1 hour and 20 minutes to complete a bicycle race. It took Johnny twice as long because he got a flat tire. How many minutes did it take Johnny to finish the race?

Name _____ Date _____

Use RDW to solve the following problems.

1. Molly baked a pie for 1 hour and 45 minutes. Then, she baked banana bread for 35 minutes less than the pie. How many minutes did it take to bake the pie and the bread?

2. A slide on the playground is $12\frac{1}{2}$ feet long. It is 3 feet 7 inches longer than the small slide. How long is the small slide?

3. The fish tank holds 8 gallons 2 quarts of water. Jeffrey poured $1\frac{3}{4}$ gallons into the empty tank. How much more water does he still need to pour into the tank to fill it?

Lesson 14: Solve multi-step word problems involving converting mixed number measurements to a single unit.

4. The candy shop puts 10 ounces of gummy bears in each box. How many boxes do they need to fill if there are $21\frac{1}{4}$ pounds of gummy bears?

5. Mom can make 10 brownies from a 12-ounce package.

   a. How many ounces of brownie mix would be needed to make 50 brownies?

   b. Extension: The brownie mix is also sold in $1\frac{1}{2}$-pound bags. How many bags would be needed to make 120 brownies?

A STORY OF UNITS   End-of-Module Assessment Task   4•7

Name _____   Date _____

1. Solve for the following conversions. Draw tape diagrams to model the equivalency.

   a. 1 gal = _____ qt

   b. 3 qt 1pt = _____ pt

2. Complete the following tables:

   a.

Pounds	Ounces
1	
2	
6	
10	
13	

   The rule for converting pounds to ounces is
   _____.

   b.

Hours	Minutes
1	
3	
7	
10	
14	

   The rule for converting hours to minutes is
   _____.

3. Answer *true* or *false* for the following statements. Explain how you know using pictures, numbers, or words.

   a. 68 ounces < 4 pounds   _____

   b. 920 minutes > 17 hours   _____

   c. 38 inches = 3 feet 2 inches   _____

EUREKA MATH   Module 7:   Exploring Measurement with Multiplication   189

4. Convert the following measurements.

   a. Express the length of a 9 kilometer trip in meters. _____

   b. Express the capacity of a 3 liter 240 milliliter container in milliliters. _____

   c. Express the length of a 3 foot 5 inch fish in inches. _____

   d. Express the length of a $2\frac{1}{4}$ hour movie in minutes. _____

   e. Express the weight of a $24\frac{3}{8}$ pound wolverine in ounces. _____

5. Find the following sums and differences. Show your work.

   a. 4 gal 2 qt + 5 gal 3 qt = _____ gal _____ qt

   b. 6 ft 2 in − 9 inches = _____ ft _____ in

   c. 3 min 34 sec + 7 min 46 sec = _____ min _____ sec

   d. 24 lb 9 oz − 3 lb 11 oz = _____ lb _____ oz

6. a. Complete the table.

Length	
yards	inches
1	
2	
3	
4	
5	
10	

b. Describe the rule for converting yards to inches.

c. How many inches are in 15 yards?

d. Jacob says that he can find the number of inches in 15 yards by tripling the number of inches in 5 yards. Does his strategy work? Why or why not?

e. A blue rope in Garret's camping backpack is 6 yards long. The blue rope is 3 times as long as a red rope. A yellow rope is 2 feet 7 inches shorter than the red rope. What is the difference in length between the blue rope and the yellow rope?

End-of-Module Assessment Task

Topics A–C

Standards Addressed

Use the four operations with whole numbers to solve problems.

4.OA.1 Interpret a multiplication equation as a comparison, e.g., interpret 35 = 5 × 7 as a statement that 35 is 5 times as many as 7 and 7 times as many as 5. Represent verbal statements of multiplicative comparisons as multiplication equations.

4.OA.2 Multiply or divide to solve word problems involving multiplicative comparison, e.g., by using drawings and equations with a symbol for the unknown number to represent the problem, distinguishing multiplicative comparison from additive comparison. (See CCSS-M Glossary, Table 2.)

4.OA.3 Solve multi-step word problems posed with whole numbers and having whole-number answers using the four operations, including problems in which remainders must be interpreted. Represent these problems using equations with a letter standing for the unknown quantity. Assess the reasonableness of answers using mental computation and estimation strategies including rounding.

Solve problems involving measurement and conversion of measurements from a larger unit to a smaller unit.

4.MD.1 Know relative sizes of measurement units within one system of units including km, m, cm; kg, g; lb, oz.; l, ml; hr, min, sec. Within a single system of measurement, express measurements in a larger unit in terms of a smaller unit. Record measurement equivalents in a two-column table. *For example, know that 1 ft is 12 times as long as 1 in. Express the length of a 4 ft snake as 48 in. Generate a conversion table for feet and inches listing the number pairs (1, 12), (2, 24), (3, 36), ...*

4.MD.2 Use the four operations to solve word problems involving distances, intervals of time, liquid volumes, masses of objects, and money, including problems involving simple fractions or decimals, and problems that require expressing measurements given in a larger unit in terms of a smaller unit. Represent measurement quantities using diagrams such as number line diagrams that feature a measurement scale.

Evaluating Student Learning Outcomes

A Progression Toward Mastery is provided to describe steps that illuminate the gradually increasing understandings that students develop *on their way to proficiency.* In this chart, this progress is presented from left (Step 1) to right (Step 4). The learning goal for students is to achieve Step 4 mastery. These steps are meant to help teachers and students identify and celebrate what the students CAN do now and what they need to work on next.

A STORY OF UNITS

End-of-Module Assessment Task 4•7

A Progression Toward Mastery				
Assessment Task Item and Standards Assessed	STEP 1 Little evidence of reasoning without a correct answer. (1 Point)	STEP 2 Evidence of some reasoning without a correct answer. (2 Points)	STEP 3 Evidence of some reasoning with a correct answer or evidence of solid reasoning with an incorrect answer. (3 Points)	STEP 4 Evidence of solid reasoning with a correct answer. (4 Points)
1 4.OA.1 4.MD.1	The student gives an incorrect answer for both parts.	The student correctly answers and models one of the two parts.	The student correctly answers both parts but has small errors in the models. OR The student correctly answers one part but draws two accurate models.	The student correctly draws a tape diagram to model each part and answers: a. 4 qt. b. 7 pt.
2 4.OA.1 4.MD.1	The student completes less than half of the problem correctly.	The student correctly completes one of the two tables with the accompanying rule.	The student correctly completes both tables but inaccurately describes the rules for one or both tables. OR The student accurately completes one table and identifies the rule for both tables.	The student correctly answers: a. 16, 32, 96, 160, and 208 ounces; identifies an accurate rule, such as *multiply by 16*. b. 60, 180, 420, 600, and 840 minutes; identifies an accurate rule, such as *multiply by 60*.
3 4.MD.1	The student correctly answers less than two parts correctly.	The student correctly answers two of the three parts, providing some reasoning for each part.	The student correctly answers all three parts and provides solid reasoning for at least two parts. OR The student correctly answers two of the three parts and provides solid reasoning for at least two parts.	The student correctly explains each answer using pictures, numbers, or words and correctly answers: a. False. b. False. c. True.

Module 7: Exploring Measurement with Multiplication

193

A STORY OF UNITS

End-of-Module Assessment Task 4•7

A Progression Toward Mastery				
4 **4.MD.1**	The student answers fewer than two parts correctly.	The student correctly answers two or three of the parts.	The student correctly answers four of the five parts. OR The student answers all parts correctly but without labeling the units.	The student correctly answers: a. 9,000 meters. b. 3,240 milliliters. c. 41 inches. d. 135 minutes. e. 390 ounces.
5 **4.MD.1** **4.MD.2**	The student answers fewer than two parts correctly.	The student correctly answers two of the four parts.	The student correctly answers three of the four parts. OR The student answers all parts correctly but does not show work.	The student correctly answers: a. 10 gal 1 qt. b. 5 ft 5 in. c. 11 min 20 sec. d. 20 lb 14 oz.
6 **4.OA.1** **4.OA.2** **4.OA.3** **4.MD.1** **4.MD.2**	The student answers fewer than three parts correctly.	The student correctly answers three of the five parts.	The student correctly answers all parts but does not provide solid reasoning or evidence of showing work in two of the five parts. OR The student correctly answers four of the five parts.	The student correctly: a. Completes the table: 36, 72, 108, 144, 180, 360 inches. b. Describes the rule, such as *multiply the number of yards by 36*. c. Solves for 540 inches in 15 yards. d. Answers *yes* and provides an accurate explanation, such as *15 yards is 3 times as much as 5 yards, so 3 × 180 inches = 540 inches*. e. Answers *14 feet 7 inches* using RDW.

Module 7:   Exploring Measurement with Multiplication

Name Jack   Date _____

1. Solve for the following conversions. Draw tape diagrams to model the equivalency.

    a. 1 gal = __4__ qt   b. 3 qt 1pt = __7__ pt

    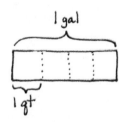   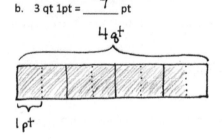

2. Complete the following tables:

   a.
Pounds	Ounces
1	16
2	32
6	96
10	160
13	208

   The rule for converting pounds to ounces is __Multiply pounds by 16__.

   b.
Hours	Minutes
1	60
3	180
7	420
10	600
14	840

   The rule for converting hours to minutes is __multiply hours by 60__.

3. Answer true or false for the following statements. Explain how you know using pictures, numbers or words.

    a. 68 ounces < 4 pounds   __false__   1 pound = 16 ounces   68 oz > 64 oz
                                                4 pounds = 64 ounces

    b. 920 minutes > 17 hours   __false__   1 hour = 60 minutes   920 min < 1,020 min
                                                17 hours = 1,020 minutes

    c. 38 inches = 3 feet 2 inches   __true__   1 foot = 12 inches   38 in = 38 in
                                                 3 feet = 36 inches
                                                 36 in + 2 in = 38 in

4. Convert the following measurements.
   a. Express the length of a 9 kilometer trip in meters. 9,000 meters

   b. Express the capacity of a 3 liter 240 milliliter container in milliliters. 3,240 mL

   c. Express the length of a 3 foot 5 inch fish in inches. 41 inches

   d. Express the length of a 2 1/4 hour movie in minutes. 135 minutes

   e. Express the weight of a 24 3/8 pound wolverine in ounces. 390 ounces

5. Find the following sums and differences. Show your work.

   a. 4 gal 2 qt + 5 gal 3 qt = __10__ gal __1__ qt

   1qt   1qt

   b. 6 ft 2 in − 9 inches = __5__ ft __5__ in

   5ft   14in

   c. 3 min 34 sec + 7 min 46 sec = __11__ min __20__ sec

   20 sec   14 sec

   d. 24 lb 9 oz − 3 lb 11 oz = __20__ lb __14__ oz

   23 lb   25 oz

6. a. Complete the table.

Yards	Inches
1	36
2	72
3	108
4	144
5	180
10	360

b. Describe the rule for converting yards to inches.

Multiply the number of yards by 36.

c. How many inches are in 15 yards?

5 yd = 180 in
10 yd = 360 in

180 in + 360 in = 540 in

There are 540 inches in 15 yards.

d. Jacob says that he can find the number of inches in 15 yards by tripling the number of inches in 5 yards. Does his strategy work? Why or why not?

Yes, 15 yards can be found multiplying 5 yards × 3, so you can multiply the number of inches in 5 yards by 3.
5 × 36 in = 180 in   180 in × 3 = 540 in

e. A blue rope in Garret's camping backpack is 6 yards long. The blue rope is 3 times as long as a red rope. A yellow rope is 2 feet 7 inches shorter than the red rope. What is the difference in length between the blue rope and the yellow rope?

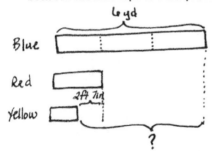

Red: 2 yd = 6 ft
6 ft − 2 ft 7 in = 3 ft 5 in
     ∧
   5 ft 12 in

Blue: 6 yd = 18 ft
18 ft − 3 ft 5 in = 14 ft 7 in
     ∧
   17 ft 12 in

The difference in length between the blue rope and the yellow rope is 14 feet 7 inches.

A STORY OF UNITS

Mathematics Curriculum

GRADE 4 • MODULE 7

Topic D
Year in Review

In Topic D, students review math concepts they learned throughout the year and create a summer folder.

In Lesson 15, students review their work with the area formula, as well as their multiplication skills, by solving for the area of composite figures. Initially introduced in Grade 3, these problems now require a deeper understanding of measurement and the area formula. Lesson 16 is a continuation of Lesson 15, asking students to draw composite figures and to solve for a determined area. To review major fluency work completed in Grade 4, students work in small groups in Lesson 17, taking turns being the teacher and delivering fluency drills to their peers. Finally, Lesson 18 reviews major vocabulary terms learned throughout Grade 4 as students play various games to further internalize these terms.

A summer folder is developed across the four lessons within this final topic. In Lesson 15, students create a take-home version of a personal white board that they can use during the lessons and at home during the summer. Each page of the Homework of these lessons includes a top half and a bottom half that have identical problems. The top half is completed for homework, ultimately becoming an answer key for the bottom half. The bottom half is placed into the student mini-personal white boards and is completed over the summer. Having already completed the top half, students can check their work by referring back to the top-half answer key. Other games, templates, and activities used in the final lessons are also included within the folder so that, on the final day of school, students go home with a folder full of activities to practice over the summer to keep their Grade 4 math skills sharp.

A Teaching Sequence Toward Mastery of a Year in Review
Objective 1: Create and determine the area of composite figures. (Lessons 15–16)
Objective 2: Practice and solidify Grade 4 fluency. (Lesson 17)
Objective 3: Practice and solidify Grade 4 vocabulary. (Lesson 18)

Lesson 15

Objective: Create and determine the area of composite figures.

Suggested Lesson Structure

- ■ Fluency Practice   (9 minutes)
- ■ Application Problem   (5 minutes)
- ■ Concept Development   (36 minutes)
- ■ Student Debrief   (10 minutes)

Total Time   **(60 minutes)**

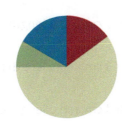

Fluency Practice (9 minutes)

- Mini-Personal White Board Set-Up   (5 minutes)
- Find the Area **4.MD.3**   (4 minutes)

Mini-Personal White Board Set-Up (5 minutes)

Materials: (S) Plastic page protector, manila folder, tape

Note: In Topic D, all homework pages are designed to become part of the take-home summer folder, created in this lesson. Therefore, students will only complete the top portion of each homework page and use the bottom portion as extra practice during the summer, inserting it into their mini-personal white boards. Although the homework does not directly reflect the work on each lesson's Problem Set, the work is directly related to the lesson and Grade 4 standards.

Today's lesson is the first of four in the Fourth Grade Project. Each lesson involves the creation of an activity page that is later placed in a take-home summer folder. (See Debrief for further explanation.) The folder will contain materials, games, and activities for student reference and practice over the summer. Students create the mini-personal white board to use during lessons and continue to use it over the summer to complete activity pages.

- Step 1: Model for students how to fold and cut a plastic page protector in half horizontally. Discard the top piece, keeping the bottom half that is closed like a pocket.
- Step 2: The pocket of the page protector becomes the mini-personal white board. It should be attached to the top of a manila folder as shown by taping the three closed sides of the page protector to the folder.

Lesson 15

Find the Area (4 minutes)

Materials: (S) Mini-personal white board

Note: This fluency activity reviews area from Module 3 and prepares students for determining the area of composite shapes in this lesson.

- T: (Project a rectangle with a width of 3 cm and a length of 10 cm.) Solve for the area of this rectangle.
- S: 30 square centimeters.
- T: (Project a rectangle with a width of 3 cm and a length of 6 cm.) Solve for the area of this rectangle.
- S: 18 square centimeters.
- T: (Project a rectangle with a width of 3 cm and a length of 16 cm.) Solve for the area of this rectangle.
- S: 48 square centimeters.

Continue with the following possible sequence:

- Rectangle with a width of 6 cm and length of 20 cm; width of 6 cm and length of 8 cm; width of 6 cm and length of 28 cm.
- Rectangle with a width of 4 cm and length of 40 cm; width of 4 cm and length of 7 cm; width of 4 cm and length of 47 cm.

Application Problem (5 minutes)

Emma's rectangular bedroom is 11 ft long and 12 ft wide. Draw and label a diagram of Emma's bedroom. How many square feet of carpet does Emma need to cover her bedroom floor?

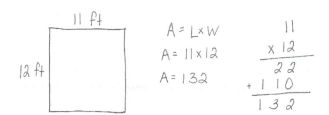

NOTES ON MULTIPLE MEANS OF ENGAGEMENT:

Give everyone a fair chance to be successful by providing appropriate scaffolds. Students may use translators, interpreters, or sentence frames to present and respond to feedback. Models shared may include concrete manipulatives.

If the pace of the lesson is a consideration, prepare presenters beforehand. The first problem may be most approachable for students working below grade level.

Note: This Application Problem reviews Grade 4's work with the area formula and two-digit by two-digit multiplication. It also serves as the lead-in to today's Concept Development. Be sure students draw models with appropriate length sides to represent the dimensions given in the problem. Have students use their mini-personal white boards to complete this problem.

Lesson 15: Create and determine the area of composite figures.

A STORY OF UNITS   Lesson 15   4•7

Concept Development (36 minutes)

Materials: (S) Problem Set

Suggested Delivery of Instruction for Solving Lesson 15's Problems

For Problems 1–6 below, students may work in pairs to solve each of the problems using the RDW approach to problem solving.

1. **Model the problem.**

Select two pairs of students who can successfully model the problem to work at the board while the other students work independently or in pairs at their seats. Review the following questions before beginning the first problem.

- Can you draw something?
- What can you draw?
- What conclusions can you make from your drawing?

As students work, circulate and reiterate the questions above. After two minutes, have the two pairs of students share *only* their labeled diagrams. For about one minute, have the demonstrating students receive and respond to feedback and questions from their peers.

2. **Calculate to solve, and write a statement.**

Allow students two minutes to complete work on the problem, sharing their work and thinking with a peer. Have the students write their equations and statements of the answer.

3. **Assess the solution.**

Give students one to two minutes to assess the solutions presented by their peers on the board, comparing the solutions to their own work. Highlight alternative methods to reach the correct solution.

Lesson 15:   Create and determine the area of composite figures.

Problem 1

Emma's rectangular bedroom is 11 ft long and 12 ft wide with an attached closet that is 4 ft by 5 ft. How many square feet of carpet does Emma need to cover both the bedroom and closet?

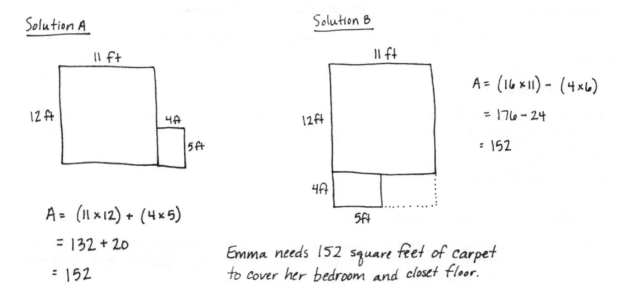

Drawing on their knowledge of solving for composite shapes in Grade 3, students may choose to solve as shown above. Solution A models solving for the two distinct areas and adding those areas together. Solution B models solving for a larger rectangle and subtracting the area not included in the floor space of the bedroom or closet.

Problem 2

To save money, Emma is no longer going to carpet her closet. In addition, she wants one 3 ft by 6 ft corner of the bedroom to be wood floor. How many square feet of carpet will she need for the bedroom now?

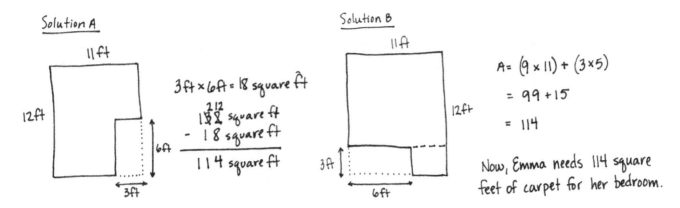

Solution A subtracts the area of the wood floor from the entire area of the bedroom. Solution B solves using just the carpeted space, dividing it into two smaller rectangles to solve. Allow students to analyze that both solution strategies are correct and the placement of the wood floor in their diagram has no effect on the answer.

Problem 3

Find the area of the figure.

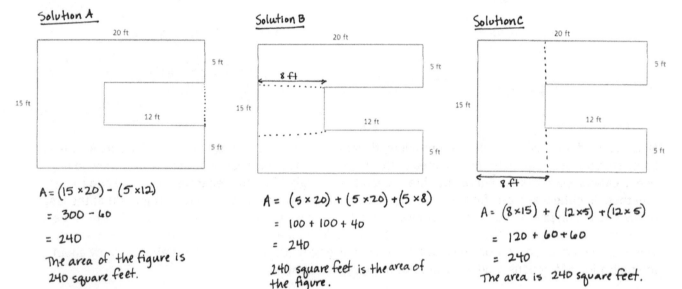

Allow students to solve using a solution that is comfortable for them. Solution A solves for the entire area of a larger rectangle and subtracts the void area. Solutions B and C partition the figure into three different smaller rectangles and find the area of the entire figure by adding the areas of the three smaller rectangles.

Problem 4

Label the sides of the figure with measurements that make sense. Find the area of the figure.

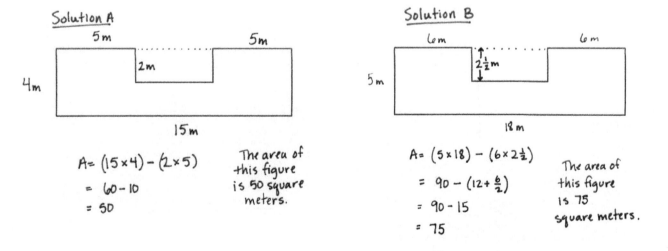

Lesson 15: Create and determine the area of composite figures.

Solution C

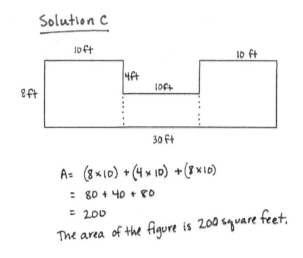

$A = (8 \times 10) + (4 \times 10) + (8 \times 10)$
$= 80 + 40 + 80$
$= 200$

The area of the figure is 200 square feet.

Solution D

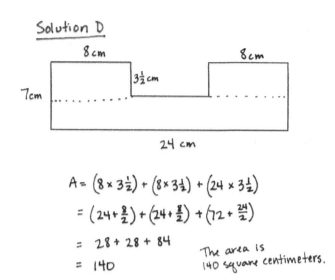

$A = (8 \times 3\frac{1}{2}) + (8 \times 3\frac{1}{2}) + (24 \times 3\frac{1}{2})$
$= (24 + \frac{8}{2}) + (24 + \frac{8}{2}) + (72 + \frac{24}{2})$
$= 28 + 28 + 84$
$= 140$

The area is 140 square centimeters.

Note: As students build and solve this problem, they may choose the strategy most comfortable for them. Some students will choose to find the area of the whole rectangle and then subtract the part they do not need. Others will break the figure into three separate rectangles, find the separate areas, and add them together. In either case, careful attention should be paid to ensuring that students recognize that the length of the single long side of the figure must be equal to the sum of the shorter opposite sides. This must be true from top to bottom, as well as from left to right. A complexity may arise, as shown in Solutions B and D, where the length of an interior side may be a fractional length. Using what students already know about multiplication of fractions, students can solve using various strategies.

Problem 5

Peterkin Park has a square fountain with a walkway around it. The fountain measures 12 feet on each side. The walkway is $3\frac{1}{2}$ feet wide. Find the area of the walkway.

Solution A

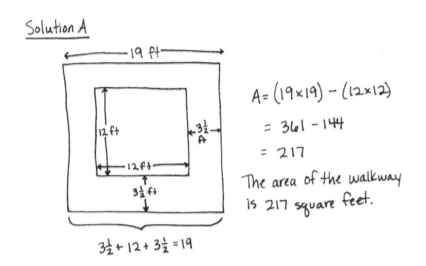

$A = (19 \times 19) - (12 \times 12)$
$= 361 - 144$
$= 217$

The area of the walkway is 217 square feet.

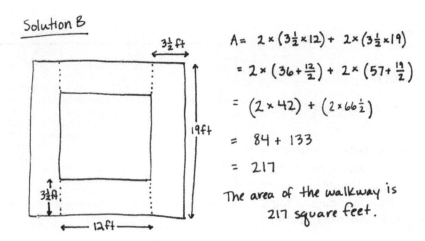

Drawing this diagram may prove difficult for some, as there are many dimensions that could be labeled. Encourage students to think about the diagram as a whole to find a solution strategy. Solution A found the area of the largest rectangle minus the area of the inner rectangle to find the area of the walkway. Solution B decomposed the walkway into 4 parts, 2 that are the same area and another 2 parts that are also the same area. Encouraging students to draw diagrams that are to scale is important for verifying if their answer is reasonable.

Problem 6

If 1 bag of gravel covers 9 square feet, how many bags of gravel will be needed to cover the entire walkway around the fountain in Peterkin Park?

Students must use division to solve this final word problem. A tape diagram allows them to see that the solution finds the number of groups not the number in each group. Students can solve using any division strategy learned in Grade 4. They must also interpret the remainder to solve correctly.

Problem Set

Please note that the Problem Set is completed as part of the Concept Development for this lesson.

A STORY OF UNITS   Lesson 15   4•7

Student Debrief (10 minutes)

Reflection (3 minutes)

Note: The Reflection replaces the Exit Ticket in Topic D.

Before the Student Debrief, instruct students to complete the Reflection pictured to the right. Reflections are replacing Exit Tickets in Topic D in order for students to have four days to think back on their learning and growth in Grade 4.

Lesson Objective: Create and determine the area of composite figures.

The Student Debrief is intended to invite reflection and active processing of the total lesson experience.

Invite students to review their reflections before going over their solutions for the Problem Set. They should check work by comparing answers with a partner before going over answers as a class. Look for misconceptions or misunderstandings that can be addressed in the Debrief. Guide students in a conversation to debrief the Problem Set and process the lesson.

Any combination of the questions below may be used to lead the discussion.

- Share your Reflection with a partner. After you have both shared, choose one more skill from the set you both notice you used today, and share your experience and progress with using that skill.
- For many word problems in Grade 4, we drew tape diagrams to model the problems. What advantage does the area model have over tape diagrams for these types of problems? How can being able to draw various models be helpful as you move into the next grades?
- In which other questions did you see each of the strategies that were used in Problem 4?
- Did you prefer using one strategy over the other? Why?
- How did Problem 3 relate to Problem 2?

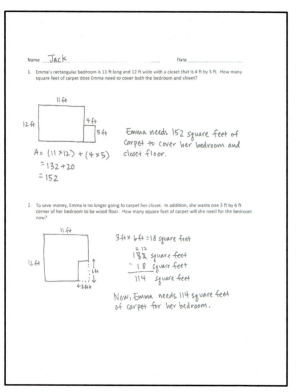

Lesson 15:   Create and determine the area of composite figures.

Model for students how their homework assignment only requires them to complete the top half of the sheet. The bottom half of each page is a duplicate of the top. By completing the top half as homework, they are creating an answer key for themselves for the summer when they can then fold the sheet in half, insert the clean problem into the mini-personal white board, and fold back the top part containing the answer. Students can then solve the problems and refer back to the answers to check their work.

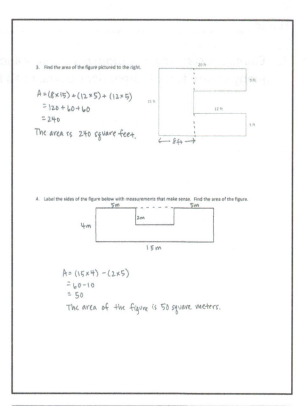
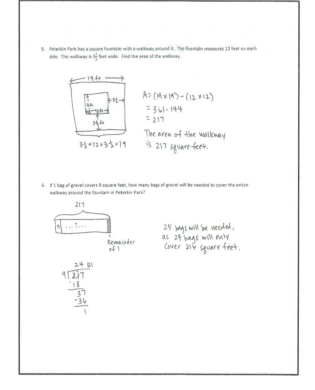

Lesson 15: Create and determine the area of composite figures.

Name _____ Date _____

1. Emma's rectangular bedroom is 11 ft long and 12 ft wide with an attached closet that is 4 ft by 5 ft. How many square feet of carpet does Emma need to cover both the bedroom and closet?

2. To save money, Emma is no longer going to carpet her closet. In addition, she wants one 3 ft by 6 ft corner of her bedroom to be wood floor. How many square feet of carpet will she need for the bedroom now?

3. Find the area of the figure pictured to the right.

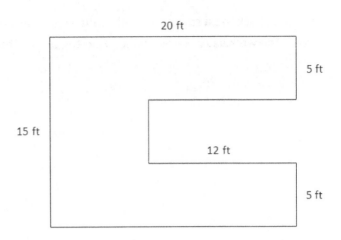

4. Label the sides of the figure below with measurements that make sense. Find the area of the figure.

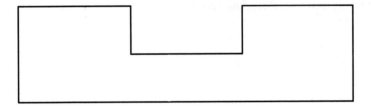

5. Peterkin Park has a square fountain with a walkway around it. The fountain measures 12 feet on each side. The walkway is $3\frac{1}{2}$ feet wide. Find the area of the walkway.

6. If 1 bag of gravel covers 9 square feet, how many bags of gravel will be needed to cover the entire walkway around the fountain in Peterkin Park?

Name _____ Date _____

In the table below are topics that you learned in Grade 4 and that were used in today's lesson.

Choose 1 topic, and describe how you were successful in using it today.

2-digit by 2-digit multiplication	Area formula	Division of 3-digit number by 1-digit number
Subtraction of multi-digit numbers	Addition of multi-digit numbers	Solving multi-step word problems

Lesson 15: Create and determine the area of composite figures.

Name _____ Date _____

For homework, complete the top portion of each page. This will become an answer key for you to refer to when completing the bottom portion as a mini-personal white board activity during the summer.

Find the area of the figure that is shaded.

1.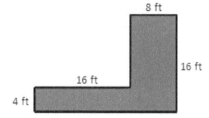

2.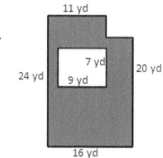

Find the area of the figure that is shaded.

1.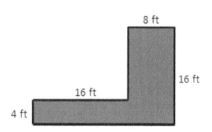

2.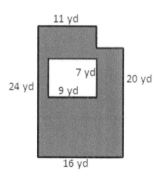

Challenge: Replace the given dimensions with different measurements, and solve again.

3. A wall is 8 feet tall and 19 feet wide. An opening 7 feet tall and 8 feet wide was cut into the wall for a doorway. Find the area of the remaining portion of the wall.

------

3. A wall is 8 feet tall and 19 feet wide. An opening 7 feet tall and 8 feet wide was cut into the wall for a doorway. Find the area of the remaining portion of the wall.

Lesson 15: Create and determine the area of composite figures.

Lesson 16

Objective: Create and determine the area of composite figures.

Suggested Lesson Structure

- ■ Fluency Practice   (9 minutes)
- ■ Concept Development   (38 minutes)
- ■ Student Debrief   (13 minutes)
- **Total Time**   **(60 minutes)**

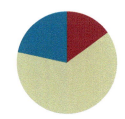

Fluency Practice (9 minutes)

- Grade 4 Core Fluency Differentiated Practice Sets 4.NBT.4   (4 minutes)
- Find the Area 4.MD.3   (5 minutes)

Grade 4 Core Fluency Differentiated Practice Sets (4 minutes)

Materials: (S) Core Fluency Practice Sets (Lesson 2 Core Fluency Practice Sets)

Note: During Module 7, each day's Fluency Practice may include an opportunity for mastery of the addition and subtraction algorithm by means of the Core Fluency Practice Sets. The process is detailed and materials are provided in Lesson 2.

Find the Area (5 minutes)

Materials: (S) Personal white board

Note: This fluency activity reviews area and solving two-digit by two-digit multiplication using the area model from Module 3. It also reviews solving for composite areas in Lesson 15.

- T: (Project a rectangle with a width of 23 cm and a length of 46 cm.) Decompose the width into tens and ones.
- S: 20 centimeters and 3 centimeters.
- T: (Draw a horizontal line decomposing the width into 20 centimeters and 3 centimeters.) Decompose the length into tens and ones.
- S: (Draw a vertical line decomposing the width into 40 centimeters and 6 centimeters.) 40 centimeters and 6 centimeters.
- T: Solve for each smaller area. Then, solve for the total area of the rectangle.
- S: 1,058 square centimeters.

Repeat the process for a rectangle with a width of 36 cm and a length of 25 cm.

A STORY OF UNITS   Lesson 16   4•7

Concept Development (38 minutes)

Materials: (S) Problem Set, protractor (Template 1 or concrete tool), centimeter ruler (Template 2 or concrete tool), large construction paper

For this lesson, students work in small groups. They use protractors and rulers to create rectangular floor plans according to the specifications given in the Problem Set. They then calculate the area of the open floor space in the floor plan.

Problem 1: Use a protractor and ruler to create a composite figure using the given specifications, and determine the area of parts of the figure.

The bedroom in Samantha's dollhouse is a rectangle 26 centimeters long and 15 centimeters wide. It has a rectangular bed that is 9 centimeters long and 6 centimeters wide. The two dressers in the room are each 2 centimeters wide. One measures 7 centimeters long, and the other measures 4 centimeters long. Create a floor plan of the bedroom containing the bed and dressers using your ruler and protractor. Find the area of the open floor space in the bedroom after the furniture is in place.

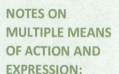

NOTES ON MULTIPLE MEANS OF ACTION AND EXPRESSION:

Scaffold Problem 1 for students working below grade level and others who may need support managing information. Provide a graphic organizer, such as the chart below, for data collection.

Item	Width	Length
bedroom		
bed		
dresser		
dresser		

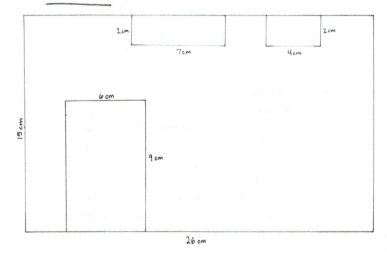

Solution A

T = 15 × 26
= 390

Total bedroom area: 390 square cm

Bed: 9 × 6 = 54
↳ 54 square cm

Dresser A: 2 × 4 = 8
↳ 8 square cm

Dresser B: 2 × 7 = 14
↳ 14 square cm

```
   26
 × 15
  130
+ 260
  390
```

A = 390 − (54 + 8 + 14)
  = 390 − 76
  = 314

The area of the open floor space is 314 square centimeters.

Lesson 16: Create and determine the area of composite figures.   215

Solution B

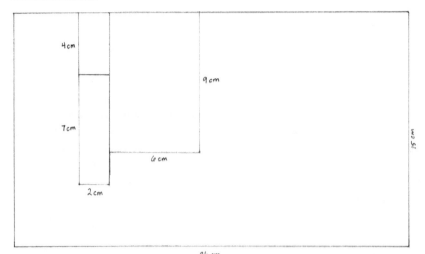

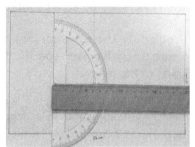
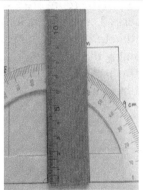

$F = (11 \times 8) - (2 \times 6)$

$= 88 - 12$

$= 76$

Total area used by furniture: 76 square centimeters

$A = (15 \times 26) - 76$

$= 390 - 76$

$= 314$

The open floor has an area of 314 square centimeters.

The sample images on the right side of the page are examples of how a ruler and a protractor can be used during the lesson to create right angles, perpendicular lines, and parallel lines to ensure students are accurately drawing rectangles. Various configurations of the furniture are acceptable. Discuss with students the real possibilities of setting up a room such as in this problem. Consider missing features like the placement of a door, window, or desk. Solution A shows a sample where the area of each piece of furniture is found, added together, and then subtracted from the total area of the room. Solution B is found by grouping the furniture. In doing so, an inner larger rectangle with the dimensions of 11 centimeters by 8 centimeters is found and the void area subtracted to find the area the furniture takes up. The area used by the furniture is then subtracted from the total area of the room. Accept reasonable solutions for solving for the amount of available floor space in the room.

A STORY OF UNITS   Lesson 16 4•7

Problem 2: Use a protractor and a ruler to create a composite figure by first using given information to determine unknown side lengths and then determining the area of part of the figure.

A model of a rectangular pool is 15 centimeters long and 10 centimeters wide. The walkway around the pool is 5 centimeters wider than the pool on each of the four sides. In one section of the walkway, there is a flowerbed that is 3 centimeters by 5 centimeters. Create a diagram of the pool area with the surrounding walkway and flowerbed. Find the area of the open walkway around the pool.

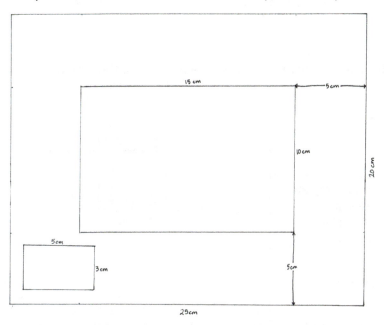

NOTES ON MULTIPLE MEANS OF ACTION AND EXPRESSION:

Provide protractor alternatives for students, if necessary. Some students may work more efficiently with large-print protractors that include a clear, moveable wand. Others may find using an angle ruler easier. Provide an appropriate adaptive ruler, such as a tactile or large-print ruler, if available and helpful.

Solution A

F = (3 × 5) + (10 × 15)
  = 15 + 150
  = 165

The pool and flowerbed take up 165 square centimeters.

A = (25 × 20) − 165
  = 500 − 165
  = 335

The walkway takes up 335 square centimeters.

Solution B

C = 2 × (20 × 5) + 2 × (5 × 15)
  = (2 × 100) + (2 × 75)
  = 200 + 150
  = 350

A = 350 − (3 × 5)
  = 350 − 15
  = 335

The open area covers 335 square centimeters.

Lesson 16:   Create and determine the area of composite figures.   217

The challenge in drawing this figure is determining the outer edges of the walkway if the pool is drawn first or determining the border of the pool if the walkway is drawn first. Although the image will fit on standard paper, consider offering students large pieces of paper so as to not have them feel restricted with the paper size. Suggest to students that they sketch the figure before drawing with a ruler and protractor. Solution A determined the area of the largest rectangle and subtracted the area of the two inner, smaller rectangles. Solution B found the area of the walkway by creating four smaller rectangles around the pool and then subtracting the area for the flowerbed.

Problem Set

Please note that the Problem Set is completed as part of the Concept Development for this lesson.

Student Debrief (13 minutes)

Reflection (3 minutes)

Before the Student Debrief, instruct students to complete the Reflection pictured to the right. Reflections are replacing the Exit Tickets in Topic D in order for students to have four days to think back on their learning and growth in Grade 4.

Lesson Objective: Create and determine the area of composite figures.

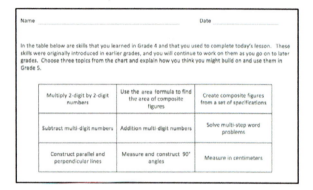

The Student Debrief is intended to invite reflection and active processing of the total lesson experience.

Invite students to review their Reflections before going over their solutions for the Problem Set. They should check work by comparing answers with a partner before going over answers as a class. Look for misconceptions or misunderstandings that can be addressed in the Debrief. Guide students in a conversation to debrief the Problem Set and process the lesson.

Any combination of the questions below may be used to lead the discussion.

- Share your Reflection with a partner. After you have both shared, choose one more skill from the set you both notice you used today, and share your experience and progress with using that skill.
- What skills from your previous work with angles and lines did you need to use today to complete the problems?
- What occupations do you think might use these ideas on a regular basis?

Name _____ Date _____

Work with your partner to create each floor plan on a separate piece of paper, as described below.

You should use a protractor and a ruler to create each floor plan and be sure each rectangle you create has two sets of parallel lines and four right angles.

Be sure to label each part of your model with the correct measurement.

1. The bedroom in Samantha's dollhouse is a rectangle 26 centimeters long and 15 centimeters wide. It has a rectangular bed that is 9 centimeters long and 6 centimeters wide. The two dressers in the room are each 2 centimeters wide. One measures 7 centimeters long, and the other measures 4 centimeters long. Create a floor plan of the bedroom containing the bed and dressers. Find the area of the open floor space in the bedroom after the furniture is in place.

2. A model of a rectangular pool is 15 centimeters long and 10 centimeters wide. The walkway around the pool is 5 centimeters wider than the pool on each of the four sides. In one section of the walkway, there is a flowerbed that is 3 centimeters by 5 centimeters. Create a diagram of the pool area with the surrounding walkway and flowerbed. Find the area of the open walkway around the pool.

Lesson 16: Create and determine the area of composite figures.

A STORY OF UNITS

Lesson 16 Reflection 4•7

Name _____ Date _____

In the table below are skills that you learned in Grade 4 and that you used to complete today's lesson. These skills were originally introduced in earlier grades, and you will continue to work on them as you go on to later grades. Choose three topics from the chart, and explain how you think you might build on and use them in Grade 5.

Multiply 2-digit by 2-digit numbers	Use the area formula to find the area of composite figures	Create composite figures from a set of specifications
Subtract multi-digit numbers	Add multi-digit numbers	Solve multi-step word problems
Construct parallel and perpendicular lines	Measure and construct 90° angles	Measure in centimeters

Lesson 16: Create and determine the area of composite figures.

Lesson 16 Homework 4•7

Name _____ Date _____

For homework, complete the top portion of each page. This will become an answer key for you to refer to when completing the bottom portion as a mini-personal white board activity during the summer.

Use a ruler and protractor to create and shade a figure according to the directions. Then, find the area of the unshaded part of the figure.

1. Draw a rectangle that is 18 cm long and 6 cm wide. Inside the rectangle, draw a smaller rectangle that is 8 cm long and 4 cm wide. Inside the smaller rectangle, draw a square that has a side length of 3 cm. Shade in the smaller rectangle, but leave the square unshaded. Find the area of the unshaded space.

----

1. Draw a rectangle that is 18 cm long and 6 cm wide. Inside the rectangle, draw a smaller rectangle that is 8 cm long and 4 cm wide. Inside the smaller rectangle, draw a square that has a side length of 3 cm. Shade in the smaller rectangle, but leave the square unshaded. Find the area of the unshaded space.

Lesson 16: Create and determine the area of composite figures.

2. Emanuel's science project display board is 42 inches long and 48 inches wide. He put a 6-inch border around the edge inside the board and placed a title in the center of the board that is 22 inches long and 6 inches wide. How many square inches of open space does Emanuel have left on his board?

- -

2. Emanuel's science project display board is 42 inches long and 48 inches wide. He put a 6-inch border around the edge inside the board and placed a title in the center of the board that is 22 inches long and 6 inches wide. How many square inches of open space does Emanuel have left on his board?

Challenge: Replace the given dimensions with different measurements, and solve again.

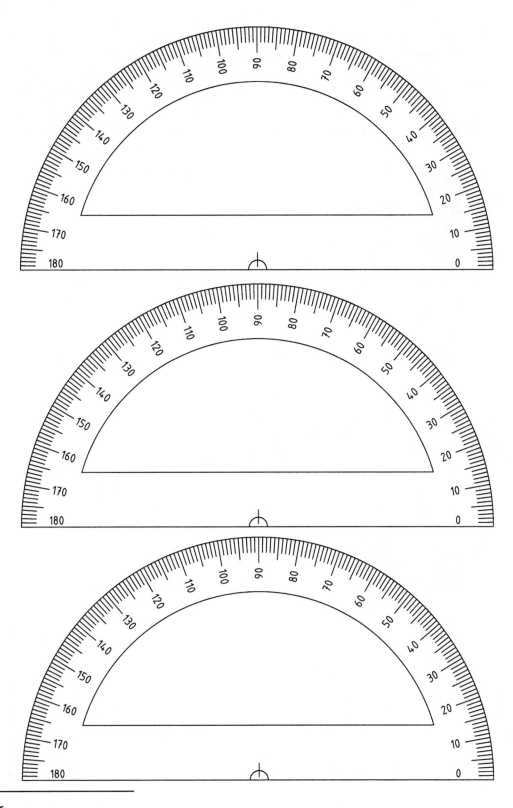

protractor

centimeter ruler

Lesson 16: Create and determine the area of composite figures.

Lesson 17

Objective: Practice and solidify Grade 4 fluency.

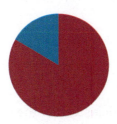

Suggested Lesson Structure

■ Fluency Practice   (50 minutes)
■ Student Debrief   (10 minutes)
  Total Time   **(60 minutes)**

Fluency Practice (50 minutes)

- Count by Equivalent Fractions   (5 minutes)
- Mixed Review Fluency   (45 minutes)

Count by Equivalent Fractions (5 minutes)

Note: Students have practiced this fluency activity throughout the year.

T: Count by threes to 30 starting at 0.
S: 0, 3, 6, 9, 12, 15, 18, 21, 24, 27, 30.

$\frac{0}{10}$	$\frac{3}{10}$	$\frac{6}{10}$	$\frac{9}{10}$	$\frac{12}{10}$	$\frac{15}{10}$	$\frac{18}{10}$	$\frac{21}{10}$	$\frac{24}{10}$	$\frac{27}{10}$	$\frac{30}{10}$
0	$\frac{3}{10}$	$\frac{6}{10}$	$\frac{9}{10}$	$\frac{12}{10}$	$\frac{15}{10}$	$\frac{18}{10}$	$\frac{21}{10}$	$\frac{24}{10}$	$\frac{27}{10}$	3
0	$\frac{3}{10}$	$\frac{6}{10}$	$\frac{9}{10}$	$1\frac{2}{10}$	$1\frac{5}{10}$	$1\frac{8}{10}$	$2\frac{1}{10}$	$2\frac{4}{10}$	$2\frac{7}{10}$	3

T: Count by 3 tenths to 30 tenths starting at 0 tenths. (Write as students count.)
S: $\frac{0}{10}, \frac{3}{10}, \frac{6}{10}, \frac{9}{10}, \frac{12}{10}, \frac{15}{10}, \frac{18}{10}, \frac{21}{10}, \frac{24}{10}, \frac{27}{10}, \frac{30}{10}$.

T: Which of these fractions is equal to a whole number?
S: 30 tenths.
T: (Point to $\frac{30}{10}$.) 30 tenths is equal to how many ones?
S: 3 ones.
T: (Beneath $\frac{30}{10}$, write 3 ones.) Count by 3 tenths again. This time, when you come to a whole number, say the whole number. (Write as students count.)
S: 0, $\frac{3}{10}, \frac{6}{10}, \frac{9}{10}, \frac{12}{10}, \frac{15}{10}, \frac{18}{10}, \frac{21}{10}, \frac{24}{10}, \frac{27}{10}$, 3.

T: (Point to $\frac{12}{10}$.) Say $\frac{12}{10}$ as a mixed number.

S: $1\frac{2}{10}$.

Continue the process for $\frac{18}{10}, \frac{21}{10}, \frac{24}{10},$ and $\frac{27}{10}$.

T: Count by 3 tenths again. This time, convert to mixed numbers or whole numbers. (Write as students count.)

S: $0, \frac{3}{10}, \frac{6}{10}, \frac{9}{10}, 1\frac{2}{10}, 1\frac{5}{10}, 1\frac{8}{10}, 2\frac{1}{10}, 2\frac{4}{10}, 2\frac{7}{10}, 3$.

Mixed Review Fluency (45 minutes)

Materials: (T) List of module titles for Modules 1–7 for the Debrief (S) Fluency cards (Template), mini-personal white board, protractor

For the rest of today's lesson students are engaged in fluency activities reviewing the major work of Grade 4. They work and play in pairs, alternating the role of teacher, using the cards provided. Students might periodically move around the room selecting different partners, or they may stay in the same grouping for the duration of this practice. Also, consider letting students select other fluency favorites based on their needs and interests.

The New Problem component of each card may be best completed after practice using the Teacher Card. The practice helps students better understand all the blanks and the movement of the teacher–student talk. They are then empowered to extend each activity. Use the mini-personal white board so that the New Problem remains usable for the summer months.

After the session, the Fluency Cards are placed in the student folders for use during the summer.

> **NOTES ON MULTIPLE MEANS OF ENGAGEMENT:**
>
> These are games that students can play with family members to maintain skills over the summer. It may be appropriate to invite parents and siblings to learn and participate, perhaps during a math or parents' night. Students may consider game partners and make adjustments accordingly. For example, if played with a younger or older sibling, games may include math appropriate for siblings. Discuss with students how to best adapt the games for their personal summer experiences.

Student Debrief (10 minutes)

Reflection (3 minutes)

Before the Student Debrief, instruct students to complete the Reflection pictured to the right. Reflections are replacing Exit Tickets in Topic D in order for students to have four days to think back on their learning and growth in Grade 4.

Lesson 17: Practice and solidify Grade 4 fluency.

Lesson Objective: Practice and solidify Grade 4 fluency.

The Student Debrief is intended to invite reflection and active processing of the total lesson experience.

Invite students to review their reflections before going over their solutions for the Problem Set. They should check work by comparing answers with a partner before going over answers as a class. Look for misconceptions or misunderstandings that can be addressed in the Debrief. Guide students in a conversation to debrief the Problem Set and process the lesson.

Any combination of the questions below may be used to lead the discussion.

- Share your Reflection with a partner. After you have both shared, talk more about ways you would like to practice this summer. What problems might you have when you try to practice?
- Do you think that, without practice, fluency can be lost? Why or why not?
- (Display a list of module titles for Modules 1–7.) We have worked hard this year and have learned many concepts in math. Let's brainstorm a list of what we have learned in math this year.
- Which of these concepts were challenging to you at first, but as you worked at them, you understood better?

Name _____ Date _____

1. What are you able to do now in math that you were not able to do at the beginning of Grade 4?

2. Which activities would you like to practice this summer in order to keep fluent or become more fluent?

3. What type of practice would help you build your fluency with these concepts?

Name _____ Date _____

1. Decimal Fraction Review: Plot and label each point on the number line below, and complete the chart. Only solve the portion above the dotted line.

Point	Unit Form	Decimal Form	Mixed Number (ones and fraction form)	How much more to get to the next whole number?
A	2 ones and 9 tenths			
B		4.4	$4\frac{4}{10}$	
C				$\frac{2}{10}$ or 0.2

----

1. Complete the chart. Create your own problem for B, and plot the point.

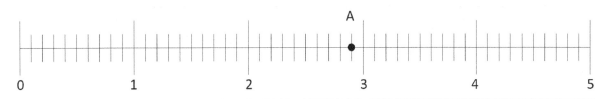

Point	Unit Form	Decimal Form	Mixed Number (ones and fraction form)	How much more to get to the next whole number?
A	2 ones and 9 tenths			
B				

Lesson 17: Practice and solidify Grade 4 fluency.

A STORY OF UNITS   Lesson 17 Homework 4•7

2. Complete the chart. The first one has been done for you. Only solve the top portion above the dotted line.

Decimal	Mixed Number	Tenths	Hundredths
3.2	$3\frac{2}{10}$	32 tenths or $\frac{32}{10}$	320 hundredths or $\frac{320}{100}$
8.6			
11.7			
4.8			

----

2. Complete the chart. Create your own problem in the last row.

Decimal	Mixed Number	Tenths	Hundredths
3.2			
8.6			
11.7			

Lesson 17: Practice and solidify Grade 4 fluency.

A STORY OF UNITS   Lesson 17 Template   4•7

Name _____   Date _____

Convert Units: Teacher Card	**New Problem**
Materials: (S) Mini-personal white board	T: (Write _____ = _____.)
T: (Write 1 m 20 cm = ____ cm.) 1 m 20 cm is how many <u>centimeters</u>?	_____ is how many _____?
S: <u>120 centimeters</u>.	S: _____.
Repeat the process with this sequence:	
1 m 80 cm = 180 cm	
3 km 249 m = 3,249 m	
4 L 71 mL = 4,071 mL	
2 kg 5 g = 2,005 g	

Add Large Numbers: Teacher Card	**New Problem**
Materials: (S) Mini-personal white board	T: (Write _____ thousands _____ ones.) On your board, write this number in standard form.
T: (Write <u>747</u> thousands <u>585</u> ones.) On your board, write this number in standard form.	S: (Write _____.)
S: (Write <u>747,585</u>.)	T: (Write _____ thousands _____ ones.) Add this number to _____ using the standard algorithm.
T: (Write <u>242</u> thousands <u>819</u> ones.) Add this number to <u>747,585</u> using the standard algorithm.	S: (_____ + _____ = _____ using the standard algorithm.)
S: (Write <u>747,585</u> + <u>242,819</u> = <u>990,404</u> using the standard algorithm.)	
Continue the process with this sequence:	
528,649 + 247,922 = 776,571	
348,587 + 629,357 = 977,944	
426,099 + 397,183 = 823,282	

fluency cards

Lesson 17:   Practice and solidify Grade 4 fluency.   231

A STORY OF UNITS

Lesson 17 Template 4•7

Subtract Large Numbers: Teacher Card

Materials: (S) Mini-personal white board

T: (Write 600 thousands.) On your board, write this number in standard form.
S: (Write 600,000.)
T: (Write 545 thousands 543 ones.) Subtract this number from 600,000 using the standard algorithm.
S: (Write 600,000 – 545,543 = 54,457 using the standard algorithm.)

Continue the process with this sequence:

400,000 – 251,559 = 148,441
700,000 – 385,476 = 314,524
600,024 – 197,088 = 402,936

New Problem

T: (Write _____ thousands .) On your board, write this number in standard form.
S: (Write _____.)
T: (Write _____ thousands _____ ones.) Subtract this number from _____ using the standard algorithm.
S: (_____ – _____ = _____ using the standard algorithm.)

Multiply Mentally: Teacher Card

Materials: (S) Mini-personal white board

T: (Write 32 × 3 = _____.) Say the multiplication sentence.
S: 32 × 3 = 96.
T: (Write 32 × 3 = 96. Below it, write 32 × 20 = _____.) Say the multiplication sentence.
S: 32 × 20 = 640.
T: (Write 32 × 20 = 640. Below it, write 32 × 23 = _____.) On your board, solve 32 × 23.
S: (Write 32 × 23 = 736.)

Repeat the process with this sequence:

42 × 2 = 84, 42 × 20 = 840, 42 × 22 = 924
31 × 4 = 124, 31 × 40 = 1,240, 31 × 44 = 1,364

New Problem

T: (Write _____ × _____ = _____.) Say the multiplication sentence.
S: _____ × _____ = _____
T: (Write _____ × _____ = _____. Below it, write _____ × _____ = _____.) Say the multiplication sentence.
S: _____ × _____ = _____.
T: (Write _____ × _____ = _____. Below it, write _____ × _____ = _____.) On your board, solve _____ × _____.
S: (Write _____ × _____ = _____.)

_____

fluency cards

Lesson 17: Practice and solidify Grade 4 fluency.

A STORY OF UNITS — Lesson 17 Template — 4•7

Divide Mentally: Teacher Card

Materials: (S) Mini-personal white board

- T: (Write 40 ÷ 2.) Write the division sentence in unit form.
- S: 4 tens ÷ 2 = 2 tens.
- T: (To the right, write 8 ÷ 2.) Write the division sentence in unit form.
- S: 8 ones ÷ 2 = 4 ones.
- T: (Write 48 ÷ 2.) Write the complete division sentence in unit form.
- S: 4 tens 8 ones ÷ 2 = 2 tens 4 ones.
- T: Say the division sentence.
- S: 48 ÷ 2 = 24.

Continue the process with this sequence:

90 ÷ 3 = 30, 3 ÷ 3 = 1, 93 ÷ 3 = 31

80 ÷ 4 = 20, 8 ÷ 4 = 2, 88 ÷ 4 = 22

180 ÷ 6 = 30, 6 ÷ 6 = 1, 186 ÷ 6 = 31

New Problem

- T: (Write _____ ÷ _____.) Write the division sentence in unit form.
- S: ____ tens ÷ ____ = ____ tens.
- T: (To the right, write _____ ÷ _____.) Write the division sentence in unit form.
- S: ____ ones ÷ ____ = ____ ones.
- T: (Write _____ ÷ _____.) Write the complete division sentence in unit form.
- S: ____ tens ____ ones ÷ ____ = ____ tens ____ ones.
- T: Say the division sentence.
- S: ____ ÷ ____ = ____.

fluency cards

Lesson 17: Practice and solidify Grade 4 fluency.

A STORY OF UNITS   Lesson 17 Template   4•7

State the Value of a Set of Coins: Teacher Card

Materials: (S) Mini-personal white board

T: (Draw 2 quarters and 4 dimes as number disks labeled 25¢ and 10¢.) What's the value of 2 quarters and 4 dimes?
S: 90¢.
T: Write 90 cents as a fraction of a dollar.
S: (Write $\frac{90}{100}$ dollar.)
T: Write 90 cents in decimal form using the dollar sign.
S: (Write $0.90.)

Continue the process with this sequence:

1 quarter 9 dimes 12 pennies = 127¢, $\frac{127}{100}$ dollar, $1.27

3 quarters 5 dimes 20 pennies = 145¢, $\frac{145}{100}$ dollar, $1.45

New Problem

T: (Draw _____ quarters and _____ dimes as number disks labeled 25¢ and 10¢.) What's the value of _____?
S: _____.
T: Write _____ cents as a fraction of a dollar.
S: (Write _____ dollar.)
T: Write _____ cents in decimal form using the dollar sign.
S: (Write $_____.)

Break Apart 180°: Teacher Card

Materials: (S) Mini-personal white board, protractor, straightedge

T: (Project a number bond with a whole of 180°. Fill in 80° for one of the parts.) On your board, complete the number bond, filling in the unknown part.
S: (Draw a number bond with a whole of 180°, and 80° and 100° as parts.)
T: Use your protractor to draw the pair of angles.
S: (Draw and label the two angles that make 180°.)

Continue the process for
120° + 60° = 180°
35° + 145° = 180°
_____ + _____ = 180°

New Problem

T: (Project a number bond with a whole of 180°. Fill in _____° for one of the parts.) On your board, complete the number bond, filling in the unknown part.
S: (Draw a number bond with a whole of 180°, and _____° and _____° as parts.)
T: Use your protractor to draw the pair of angles.
S: (Draw and label the two angles that make 180°.)

fluency cards

Lesson 17:   Practice and solidify Grade 4 fluency.

EUREKA MATH

Lesson 18

Objective: Practice and solidify Grade 4 vocabulary.

Suggested Lesson Structure

■ Fluency Practice   (8 minutes)
■ Concept Development   (42 minutes)
■ Student Debrief   (10 minutes)
    Total Time   **(60 minutes)**

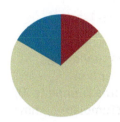

Fluency Practice (8 minutes)

- Grade 4 Core Fluency Differentiated Practice Sets   **4.NBT.4**   (4 minutes)
- Draw and Identify Geometric Terms   **4.G.1**   (4 minutes)

Grade 4 Core Fluency Differentiated Practice Sets (4 minutes)

Materials: (S) Core Fluency Practice Sets (Lesson 2 Core Fluency Practice Sets)

Note: During Module 7, each day's Fluency Practice may include an opportunity for mastery of the addition and subtraction algorithm by means of the Core Fluency Practice Sets. The process is detailed and materials are provided in Lesson 2. It is recommended these sets be sent home in the summer folder.

Draw and Identify Geometric Terms (4 minutes)

Materials: (S) Personal white board, protractor, ruler

Note: This fluency activity reviews Module 4 and prepares students for using geometric terms in today's lesson.

- T: Use your protractor and ruler to draw a right, isosceles triangle.
- S: (Draw as shown to the right, though student pictures may vary.)
- T: Label vertices to identify the right angle as $\angle ABC$.
- S: (Label as shown to the right.)
- T: $\overline{AB}$ and $\overline{BC}$ are what types of lines?
- S: Perpendicular lines.
- T: Use your protractor and ruler to draw a rectangle $DEFG$.
- S: (Draw as shown to the right, though student pictures may vary.)

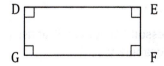

A STORY OF UNITS

T: What type of lines are $\overline{DE}$ and $\overline{FG}$?
S: Parallel lines.
T: Identify another pair of parallel lines.
S: $\overline{EF}$ and $\overline{DG}$.

Concept Development (42 minutes)

Materials: (S) 2 small envelopes or baggies containing cardstock cutouts of game descriptions (Template 1) and vocabulary cards (Template 3), math bingo card on cardstock (Template 2), timer (1 per group), summer folder

For the rest of today's lesson, students play vocabulary games reviewing the major work of Grade 4. Consider opening the lesson with a game of bingo with the whole class and then having them play either bingo or one of the other games in pairs or groups of four, alternating the role of caller, using the cards provided. As was done yesterday, students might periodically move around the room selecting different partners and playing one of the four games, or they might stay in the same grouping for the duration of this practice.

After the session, store the instructions for the games and all materials in the summer folders for home use.

Problem Set

Please note that the Problem Set for Lesson 18 is math bingo and other games students play in class.

NOTES ON MULTIPLE MEANS OF ENGAGEMENT:

Like yesterday's fluency activities, these are games that students can play with family members to maintain skills over the summer. It may be appropriate to invite parents and siblings to learn and participate, perhaps at a math or parents' night. Students may consider their game partner and make adjustments accordingly. For example, if played with a younger or older sibling, games may include math appropriate for siblings. Discuss with students how to best adapt the games for their personal summer experiences.

NOTES ON MULTIPLE MEANS OF REPRESENTATION:

To accommodate English language learners during games such as Bingo and Concentration, rather than using all 24 words during the games, consider omitting some words to reduce the amount of reading.

Student Debrief (10 minutes)

Reflection (3 minutes)

Before the Student Debrief, instruct students to complete the Reflection pictured to the right. Reflections are replacing the Exit Tickets in Topic D in order for students to have four days to think back on their learning and growth in Grade 4.

Lesson Objective: Practice and solidify Grade 4 vocabulary.

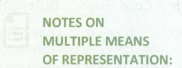

The Student Debrief is intended to invite reflection and active processing of the total lesson experience.

Invite students to review their Reflections before going over their solutions for the Problem Set. They should check work by comparing answers with a partner before going over answers as a class. Look for misconceptions or misunderstandings that can be addressed in the Debrief. Guide students in a conversation to debrief the Problem Set and process the lesson.

Any combination of the questions below may be used to lead the discussion.

- Share your Reflection with a partner. After you have both shared, talk more about ways you might practice this summer and how to overcome difficulties with practicing.
- Which games did you most enjoy? Who might you play those games with during the summer?
- Which games were the most challenging? Did you enjoy the challenge?
- How might you modify the games to play with family and friends?
- How does vocabulary help you to communicate with the people who care about you, about your education, and about what happens in school?

Name _____ Date _____

1. Why do you think vocabulary was such an important part of fourth-grade math? How does vocabulary help you in math?

2. Which vocabulary terms do you know well, and which would you like to improve upon?

Bingo:

1. Players write a vocabulary term in each box of the math bingo game. Each term should be used only once. The box that says *Math Bingo* is a free space.
2. Players place the filled-in math bingo template in their mini-personal white boards.
3. One person is the caller and reads the definition on a vocabulary card.
4. Players cross off (or cover) the term that matches the definition.
5. *Bingo!* is called when 5 vocabulary terms in a row are crossed off diagonally, vertically, or horizontally. The free space counts as 1 box toward the needed 5 vocabulary terms.
6. The first player to have 5 in a row reads each crossed off word, states the definition, and gives a description or an example of each word. If all words are reasonably explained as determined by the caller, the player is declared the winner.

Math Jeopardy:

Structure: Teams or partnerships. Callers should prepare the game in advance.

1. The definitions are sorted into labeled columns by a caller: units, lines and angles, the four operations, and geometric shapes.
2. The first term directly below the heading has a value of $100, the next $200, and so on. The caller should make an effort to order the questions from easiest to hardest.
3. Player 1 chooses a column and a dollar value, for example, "I choose geometry terms for $100." The caller reads, "The answer is…"
4. The players say the matching question, for example, "What is a quadrilateral?"
5. The first person to correctly state the question wins the dollar value for that card.
6. Play continues until all cards are used.
7. The player with the highest dollar value wins.

Concentration:

Structure: Teams or partnerships.

1. Create an array of all the cards face down.
2. Players take turns flipping over pairs of cards to find a match. A match is a vocabulary term and its definition. Cards keep their precise location in the array if not matched. Remaining cards are not reconfigured into a new array.
3. After all cards are matched, the player with the most pairs is the winner.

Math Pictionary:

Structure: Teams or partnerships.

1. A timer is set for 1 minute.
2. A vocabulary term is chosen from a bag by a player from Team 1, who draws an example as quickly as possible.
3. The player's teammate(s) tries to guess the vocabulary term. When the term is guessed, a new term is chosen by the same player. The process is repeated as many times as possible within the minute. Terms not guessed when the timer sounds go back in the bag.
4. A player from Team 2 repeats the process.
5. Teams count the number of words guessed. The team with the most words is the winner.

game descriptions

Lesson 18: Practice and solidify Grade 4 vocabulary.

A STORY OF UNITS

Lesson 18 Template 2   4•7

		Math BINGO!		

		Math BINGO!		

math bingo

Lesson 18: Practice and solidify Grade 4 vocabulary.

A metric unit of measure equivalent to 1,000 grams.	A whole number greater than 1 whose only factors are 1 and itself.	An angle measuring less than 90 degrees.	Lines that intersect and form a 90° angle.
A whole number plus a fraction.	An angle that turns through $\frac{1}{360}$ of a circle.	The bottom number in a fraction that tells the number of equal parts in the whole.	A customary unit of measurement for liquid volume equivalent to 4 quarts.
A customary unit of measurement for liquid volume equivalent to 2 pints.	The answer to a multiplication problem.	A number leftover that can't be divided into equal groups.	A line through a figure such that when the figure is folded along the line, two halves are created that match up exactly.
Two lines in a plane that never intersect.	A triangle with at least two equal sides.	A whole number having three or more distinct factors.	A closed figure with 4 straight sides and 4 angles.
An angle measuring 90 degrees.	An angle with a measure greater than 90 degrees but less than 180 degrees.	Lines that contain at least 1 point in common.	A tool used to measure and draw angles.
The top number in a fraction that tells how many parts of the whole are selected.	A triangle that contains one 90-degree angle.	This special angle measures 180 degrees.	A closed figure with 3 straight sides of equal length and 3 equal angles.

vocabulary cards (page 1)

Kilogram	Prime Number	Acute Angle	Perpendicular Lines
Mixed Number	One-Degree Angle	Denominator	Gallon
Quart	Product	Remainder	Line of Symmetry
Parallel Lines	Isosceles Triangle	Composite Number	Quadrilateral
Right Angle	Obtuse Angle	Intersecting Lines	Protractor
Numerator	Right Triangle	Straight Angle	Equilateral Triangle

vocabulary cards (page 2)

Answer Key

Eureka Math
Grade 4
Module 7

Special thanks go to the Gordon A. Cain Center and to the Department of Mathematics at Louisiana State University for their support in the development of *Eureka Math*.

For a free *Eureka Math* Teacher Resource Pack, Parent Tip Sheets, and more please visit www.Eureka.tools

Published by the non-profit Great Minds

Copyright © 2015 Great Minds. No part of this work may be reproduced, sold, or commercialized, in whole or in part, without written permission from Great Minds. Non-commercial use is licensed pursuant to a Creative Commons Attribution-NonCommercial-ShareAlike 4.0 license; for more information, go to http://greatminds.net/maps/math/copyright. "Great Minds" and "Eureka Math" are registered trademarks of Great Minds.

Printed in the U.S.A.
This book may be purchased from the publisher at eureka-math.org
10 9 8 7 6 5 4 3 2

A STORY OF UNITS

Mathematics Curriculum

GRADE 4 • MODULE 7

Answer Key
GRADE 4 • MODULE 7
Exploring Measurement with Multiplication

Lesson 1

Sprint

Side A

1. $0.01
2. $0.02
3. $0.03
4. $0.08
5. $0.80
6. $0.70
7. $0.60
8. $0.20
9. $0.01
10. $0.10
11. $0.02
12. $0.20
13. $0.03
14. $0.30
15. $0.90
16. $0.07
17. $0.80
18. $0.04
19. $0.60
20. $0.08
21. $0.70
22. $0.09
23. $0.06
24. $0.50
25. $0.05
26. $0.11
27. $0.12
28. $0.17
29. $0.45
30. $0.63
31. $0.63
32. $0.97
33. $0.25
34. $0.50
35. $0.75
36. $0.53
37. $0.28
38. $0.78
39. $0.70
40. $0.35
41. $0.85
42. $0.65
43. $0.95
44. $0.93

Side B

1. $0.02
2. $0.03
3. $0.04
4. $0.09
5. $0.90
6. $0.80
7. $0.70
8. $0.30
9. $0.01
10. $0.10
11. $0.02
12. $0.20
13. $0.03
14. $0.30
15. $0.80
16. $0.06
17. $0.70
18. $0.09
19. $0.50
20. $0.07
21. $0.90
22. $0.08
23. $0.05
24. $0.60
25. $0.04
26. $0.11
27. $0.12
28. $0.18
29. $0.54
30. $0.74
31. $0.74
32. $0.86
33. $0.25
34. $0.50
35. $0.75
36. $0.54
37. $0.29
38. $0.79
39. $0.80
40. $0.45
41. $0.95
42. $0.75
43. $0.85
44. $0.94

Lesson 1 Answer Key 4•7

Practice Sheet

a. 16, 32, 48, 64, 80, 96, 112, 128, 144, 160; multiply the number of pounds times 16.

b. 3, 6, 9, 12, 15, 18, 21, 24, 27, 30; multiply the number of yards times 3.

c. 12, 24, 36, 48, 60, 72, 84, 96, 108, 120; multiply the number of feet times 12.

Problem Set

1. 32 1-ounce weights
2. 13 1-ounce weights
3. 84 ounces
4.
   a. 16, 48, 112, 160, 272; multiply the number of pounds times 16.
   b. 12, 24, 60, 120, 180; multiply the number of feet times 12.
   c. 3, 6, 12, 30, 42; multiply the number of yards times 3.

5.
   a. 37
   b. 142
   c. 16
   d. 38
   e. 442
   f. 63
   g. 229
   h. 204

6.
   a. False; 2 kg > 1,600 g
   b. False; 12 ft < 150 in
   c. True

Exit Ticket

1.
   a. 96
   b. 14
   c. 231

2.
   a. False; 3 lb > 47 ounces
   b. True

Homework

1. a. 3, 6, 9, 15, 30
   b. 12, 24, 60, 120, 180
   c. 36, 108, 216, 360, 432
2. a. 74
   b. 334
   c. 14
   d. 40
   e. 206
   f. 34
   g. 47
   h. 204
3. 240 inches
4. 16, 32, 64, 160, 192
5. 114 ounces
6. a. True
   b. False; 10 yd < 361 in
   c. False; 10 liters = 10,000 mL

Lesson 2

Core Fluency Practice Set A

Part 1
1. 15,413
2. 51,294
3. 88,001
4. 683,874
5. 483,391
6. 932,476

Part 2
1. 15,413
2. 51,294
3. 88,001
4. 683,874
5. 483,391
6. 932,476

Core Fluency Practice Set B

Part 1
1. 2,193
2. 18,016
3. 9,502
4. 400,685

Part 2
1. 2,193
2. 18,016
3. 9,502
4. 400,685

Core Fluency Practice Set C

Part 1
1. 2,418
2. 22,148
3. 25,063
4. 375,707

Part 2
1. 2,418
2. 22,148
3. 25,063
4. 375,707

Lesson 2 Answer Key 4•7

Core Fluency Practice Set D

Part 1
1. 18,991
2. 525
3. 793,672
4. 811,939
5. 632,594
6. 591,071

Part 2
1. 18,991
2. 525
3. 793,672
4. 811,939
5. 632,594
6. 591,071

Practice Sheet

a. 4, 8, 12, 16, 20, 24, 28, 32, 36, 40; multiply the number of gallons times 4.

b. 2, 4, 6, 8, 10, 12, 14, 16, 18, 20; multiply the number of quarts times 2.

c. 2, 4, 6, 8, 10, 12, 14, 16, 18, 20; multiply the number of pints times 2.

d. 8; 4; 16

Problem Set

1. 6 pints
2. 6 quarts
3. 4,000 mL
4. a. 4, 12, 20, 40, 52; multiply the number of gallons times 4.
   b. 2, 4, 12, 20, 32; multiply the number of quarts times 2.
5. a. 34
   b. 62
   c. 18
   d. 54
   e. 214
   f. 520
6. a. False; 1 gallon > 3 quarts
   b. True
   c. False; 15 pints < 2 gallons
7. 2,500 doses
8. Moore family
9. 144 cups

Exit Ticket

1. 4, 8, 16
2. No; answers will vary.

Homework

1. 48 cups
2. 6 quarts (or equivalent)
3. 1,720 mL (or equivalent)
4. a. 4, 8, 16, 48, 60
   b. 2, 4, 12, 20, 32
5. a. 27
   b. 50
   c. 11
   d. 58
   e. 140
   f. 444
6. Answers will vary.
7. a. False; 2 quarts > 3 pints
   b. True
   c. True
8. 6 pints; answers will vary.
9. 12 cups

Lesson 3

Practice Sheet

a. 60, 120, 180, 240, 300, 360, 420, 480, 540, 600; multiply the number of minutes times 60.

b. 60, 120, 180, 240, 300, 360, 420, 480, 540, 600; multiply the number of hours times 60.

c. 24, 48, 72, 96, 120, 144, 168, 192, 216, 240; multiply the number of days times 24.

Problem Set

1. 120 minutes
2. Under 360 minutes
3. a. 60, 180, 360, 600, 900; multiply by the number of hours times 60.
   b. 24, 48, 120, 168, 240; multiply the number of days times 24.
4. a. 570
   b. 465
   c. 236
   d. 1,347
   e. 331
   f. 1,385
5. Explanations will vary.
6. 883 seconds
7. 744 hours

Exit Ticket

1,324 minutes

Homework

1. 180 minutes
2. 300 minutes
3. a. 60, 120, 300, 540, 720; multiply the number of hours times 60.
   b. 24, 72, 144, 192, 480; multiply the number of days times 24.
4. a. 630
   b. 375
   c. 116
   d. 225
   e. 573
   f. 1,025
5. Explanations will vary.
6. 516 seconds
7. 264 hours

Lesson 4

Problem Set

1. 240 minutes
2. 112 ounces
3. 36 feet
4. 66,000 milliliters
5. 86 ounces

Exit Ticket

8 ounces

Homework

1. 360 minutes
2. 56 ounces
3. 1,350 milliliters
4. 12 feet
5. 14 boxes
6. a. 45 quarts (or equivalent)
   b. No; answers will vary.

Lesson 5

Sprint

Side A

1. 1,000
2. 2,000
3. 3,000
4. 7,000
5. 5,000
6. 100
7. 200
8. 300
9. 900
10. 600
11. 3
12. 6
13. 9
14. 30
15. 15
16. 12
17. 24
18. 36
19. 120
20. 48
21. 9,000
22. 4,000
23. 6,000
24. 500
25. 700
26. 400
27. 800
28. 12
29. 24
30. 18
31. 27
32. 60
33. 72
34. 1
35. 8
36. 1
37. 6
38. 1
39. 8
40. 1
41. 6
42. 96
43. 7
44. 108

Side B

1. 100
2. 200
3. 300
4. 700
5. 500
6. 1,000
7. 2,000
8. 3,000
9. 9,000
10. 6,000
11. 3
12. 6
13. 9
14. 15
15. 30
16. 12
17. 24
18. 36
19. 120
20. 48
21. 900
22. 400
23. 600
24. 5,000
25. 7,000
26. 4,000
27. 8,000
28. 18
29. 27
30. 12
31. 24
32. 60
33. 72
34. 1
35. 8
36. 1
37. 6
38. 1
39. 9
40. 1
41. 7
42. 108
43. 6
44. 96

A STORY OF UNITS

Lesson 5 Answer Key 4•7

Problem Set

1. a. Tape diagram labeled; 11 feet 7 inches
   b. Answers will vary.
2. Answers will vary; 29 pounds 14 ounces

Exit Ticket

1,340 yards

Homework

1. 24 cups
2. 36 minutes
3. 12 ounces
4. a. Tape diagram labeled; 19 feet 2 inches
   b. Answers will vary.
5. Answers will vary; 21 pounds 8 ounces

Module 7: Exploring Measurement with Multiplication

Lesson 6

Problem Set

1. a. 1
    b. 3
    c. 3
    d. 4, 3
    e. 1
    f. 3
    g. 1
    h. 4, 1

2. a. 7, 2
    b. 14, 2
    c. 8, 7
    d. 4, 2
    e. 17, 2
    f. 10, 0

3. 1 quart 1 cup, or 5 cups

4. a. 44 cups (or equivalent)
    b. 4 cups

Exit Ticket

1. a. 11, 1
    b. 3, 2

2. 3 quarts

Homework

1. a. 2
    b. 2
    c. 1
    d. 2, 2
    e. 1
    f. 4
    g. 1
    h. 3, 1

2. a. 5, 1
    b. 18, 1
    c. 6, 7
    d. 6, 5
    e. 14, 3
    f. 14, 2

3. 1 gallon 2 quarts, or 6 quarts

4. a. 49 cups (or equivalent)
    b. 15 cups

Lesson 7

Problem Set

1. a. 1
   b. 4
   c. 2
   d. 7, 2
   e. 1
   f. 1, 3
   g. 4
   h. 4, 4

2. a. 6, 1
   b. 10, 1
   c. 3, 2
   d. 3, 2
   e. 7, 1
   f. 8, 3
   g. 33, 8
   h. 1, 3

3. 2 ft 8 in, or 32 in
4. 14 ft 1 in, or 169 in
5. a. 5 ft 10 in, or 70 in
   b. 17 ft 6 in, or 210 in

Exit Ticket

1. 5
2. 5, 2
3. 8
4. 4, 2

Homework

1. a. 3
   b. 1, 2
   c. 1, 1
   d. 4, 2
   e. 1
   f. 1, 2
   g. 10
   h. 1, 6

2. a. 5, 1
   b. 8, 0
   c. 4, 2
   d. 1, 2
   e. 8, 1
   f. 12, 2
   g. 31, 8
   h. 4, 3

3. 15 ft 9 in, or 189 in
4. 8 ft 9 in, or 105 in
5. a. 12 ft 6 in, or 150 in
   b. Length is equal to width.

Lesson 8

Problem Set

1. a. 1
   b. 2
   c. 3
   d. 11, 12
   e. 4, 2
   f. 40, 2
   g. 24, 4
   h. 112, 15

2. 11 pounds 7 ounces, or 183 ounces
3. 1 pound 2 ounces
4. a. 9 pounds 10 ounces, or 154 ounces
   b. 11 pounds 13 ounces, or 189 ounces

Exit Ticket

1. 5, 0
2. 16, 2
3. 4, 8
4. 6, 14

Homework

1. a. 1
   b. 2
   c. 5
   d. 11, 8
   e. 6, 1
   f. 28, 1
   g. 22, 2
   h. 72, 14

2. 4 pounds 11 ounces, or 75 ounces
3. 3 ounces
4. a. 6 pounds 9 ounces, or 105 ounces
   b. 17 pounds 5 ounces, or 277 ounces
   c. 6 pounds 11 ounces, or 107 ounces

Lesson 9

Problem Set

1. a. 1
   b. 2
   c. 48
   d. 3, 48
   e. 1
   f. 2, 15

2. a. 4, 10
   b. 9, 10
   c. 2, 25
   d. 2, 54
   e. 6, 7
   f. 16, 50

3. 2 minutes 23 seconds, or 143 seconds

4. a. No; explanations will vary.
   b. 21 minutes

Exit Ticket

1. 2, 50
2. 7, 20
3. 10, 34
4. 1, 27

Homework

1. a. 1
   b. 3
   c. 27
   d. 2, 27
   e. 1
   f. 4, 45

2. a. 6, 5
   b. 9, 10
   c. 3, 26
   d. 2, 38
   e. 4, 9
   f. 16, 39

3. 5 minutes 7 seconds, or 307 seconds

4. a. No; explanations will vary.
   b. 4 hours 28 minutes, or 268 minutes;
      2 hours 32 minutes, or 152 minutes

Lesson 10

Problem Set

1. 12 hours 40 minutes, or 760 minutes
2. 23 gallons 1 quart, or 93 quarts
3. 30 pounds 13 ounces, or 493 ounces
4. 2 feet 11 inches, or 35 inches

Exit Ticket

2 hours 35 minutes, 155 minutes

Homework

1. 1 cup
2. 5 quarts 1 cup, or 21 cups
3. 5 hours 25 minutes, or 325 minutes
4. 2 feet 6 inches
5. 155 pounds 12 ounces

Lesson 11

Problem Set

1. 4 hours 10 minutes, or 250 minutes
2. 99
3. 154
4. Yes

Exit Ticket

55 minutes

Homework

1. 3 hours 43 minutes, or 223 minutes
2. 25 feet 3 inches, or 303 inches
3. 80
4. 1 hour 31 minutes, or 91 minutes
5. Yes

Lesson 12

Problem Set

1. Tape diagram drawn
   a. 1
   b. 2
   c. 3
2. Tape diagram drawn to show equivalence
3. Tape diagram drawn to show equivalence
4. Tape diagram drawn to show equivalence
5. a. 1
   b. 6, 6
   c. 3, 3
   d. 9, 9
   e. 4, 4
   f. 8, 8

6. a. 4
   b. 14
   c. 10
   d. 31
   e. 18
   f. 78
   g. 15
   h. 75

Exit Ticket

1. a. 6, 6
   b. 9, 9
2. a. 4
   b. 23

Homework

1. Tape diagram drawn to show equivalence
2. Tape diagram drawn to show equivalence
3. Tape diagram drawn to show equivalence
4.
   a. 6
   b. 3, 3
   c. 2, 2
   d. 4, 4
   e. 8, 8
   f. 10, 10

5.
   a. 8
   b. 10
   c. 14
   d. 21
   e. 75
   f. 88
   g. 30
   h. 69
   i. 116
   j. 94

Lesson 13

Problem Set

1. a. 1
   b. 8, 8
   c. 4, 4
   d. 12, 12
   e. 2, 2
   f. 6, 6
2. Tape diagram drawn to show equivalence
3. a. 1
   b. 30, 30
   c. 15, 15
4. Tape diagram drawn to show equivalence

5. a. 18
   b. 54
   c. 92
   d. 88
   e. 75
   f. 210
   g. 135
   h. 330
   i. 10
   j. 23
   k. 18
   l. 27
   m. 69
   n. 100

Exit Ticket

1. Tape diagram drawn to show equivalence
2. a. 20
   b. 165
   c. 66
   d. 46

Homework

1. a. 1
    b. 8, 8
    c. 4, 4
    d. 12, 12
    e. 2, 2
    f. 10, 10
2. Tape diagram drawn to show equivalence
3. a. 1
    b. 30, 30
    c. 15, 15
    d. 20, 20
4. Tape diagram drawn to show equivalence

5. a. 36
    b. 78
    c. 108
    d. 66
    e. 105
    f. 270
    g. 225
    h. 320
    i. 14
    j. 19
    k. 17
    l. 11
    m. 75
    n. 118

Lesson 14

Problem Set

1. 210 minutes
2. 69 inches
3. 10 gallons 3 quarts, or 43 quarts
4. 80 inches
5. 72
6. a. 90 ounces
   b. 3

Exit Ticket

160 minutes

Homework

1. 175 minutes
2. 8 feet 11 inches, or 107 inches
3. 6 gallons 3 quarts, or 27 quarts
4. 34
5. a. 60 ounces
   b. 6

Lesson 15

Problem Set

1. 152 square feet
2. 114 square feet
3. 240 square feet
4. Answers will vary.
5. 217 square feet
6. 25

Reflection

Answers will vary.

Homework

1. 192 square feet
2. 301 square yards
3. 96 square feet

Lesson 16

Problem Set

1. 314 square centimeters
2. 335 square centimeters

Reflection

Answers will vary.

Homework

1. 85 square centimeters
2. 948 square inches

Lesson 17

Reflection

1. Answers will vary.
2. Answers will vary.
3. Answers will vary.

Homework

1. A. Point accurately plotted on number line; 2.9; $2\frac{9}{10}$; $\frac{1}{10}$ or 0.1

   B. Point accurately plotted on number line; 4 ones and 4 tenths; $\frac{6}{10}$ or 0.6

   C. Answers will vary.

2. Answer provided

   $8\frac{6}{10}$; 86 tenths or $\frac{86}{10}$; 860 hundredths or $\frac{860}{100}$

   $11\frac{7}{10}$; 117 tenths or $\frac{117}{10}$; 1170 hundredths or $\frac{1170}{100}$

   $4\frac{8}{10}$; 48 tenths or $\frac{48}{10}$; 480 hundredths or $\frac{480}{100}$

Template

Convert Units: 120

Add Large Numbers: Answers provided

Subtract Large Numbers: Answers provided

Multiply Mentally: 96; 640; 736

Divide Mentally: Answers provided

State the Value of a Set of Coins: Answers provided

Break Apart 180°: Answers provided

Lesson 18

Reflection

1. Answers will vary.

2. Answers will vary.